William E. Russey, Hans F. Ebel,
Claus Bliefert

How to Write
a Successful Science Thesis

William E. Russey, Hans F. Ebel,
Claus Bliefert

How to Write
a Successful Science Thesis

The Concise Guide for Students

WILEY-VCH Verlag GmbH & Co. KGaA

Authors

Dr. William E. Russey
3508 Cold Springs Road
Huntingdon, PA 16652
USA
russey@juniata.edu

Dr. Hans Friedrich Ebel
Im Kantelacker 15
64646 Heppenheim
Germany
ebel-heppenheim@t-online.de

Prof. Dr. Claus Bliefert
Meisenstraße 60
48624 Schöppingen
Germany
bliefert@fh-muenster.de

Library of Congress Card No.: applied for

British Library Cataloguing-in-Publication Data
A catalogue record for this book is available from the British Library

Bibliographic information published by Die Deutsche Bibliothek
Die Deutsche Bibliothek lists this publication in the Deutsche Nationalbibliografie; detailed bibliographic data is available in the Internet at <http://dnb.ddb.de>

Printed in the Federal Republic of Germany
Printed on acid-free paper

Typesetting Claus Bliefert, Schöppingen
Printing betz-druck GmbH, Darmstadt
Binding Litges & Dopf Buchbinderei GmbH, Heppenheim

ISBN-13: 978-3-527-31298-6
ISBN-10: 3-527-31298-6

Preface

It is both humbling and gratifying to know that, since 1987, our book

Ebel, H. F., Bliefert, C., and Russey, W. E.
The Art of Scientific Writing. Weinheim: VCH

has been accepted by so many as an authoritative and comprehensive guide to the general field of scientific communication. A surprising number of reviewers have in fact referred to it as "a handbook" on the subject. (A new, completely revised and updated edition was released in early 2004.) Several critics have even ventured to recommend it as a suitable basis upon which to develop a university-level course on the subject. Should

The Art ... therefore be characterized as a textbook? Perhaps, but in a sense it has the distinct disadvantage of presenting *too much* information.

Something that in our judgement has heretofore been lacking is a concise—and at the same time user-friendly!—introductory guide to preparing the sort of formal report that serves as the culmination of a science student's first serious personal encounter with meaningful research. This document is most often referred to, whether at the graduate or the undergraduate level, as a *thesis.*

Our goal in preparing the present slender volume has been to create just such an introduction. We took as our starting point a similar (and generally well-received) book published some years ago in German by the Ebel/Bliefert core of our triumvirate (*Diplom- und Doktorarbeit: Anleitungen für den naturwissenschaftlich-technischen Nachwuchs*, currently available as a third edition from Wiley–VCH).

We concentrate this time strictly on the basics, attempting to cover no more but also no less. Our target audience is primarily aspiring scientists still wending their way through the world of academe. We have structured the material in a way that should lend itself to either a classroom environment or conscientious self-study, with perhaps just a hint of programmed-learning character. The information has been subdivided into twenty-one "units", each providing guidance within a single general context. A typical unit could easily be given a title of the "How to ..." sort: e.g., "How to prepare ..." or "How to structure ..." or "How to go about ...". Each opens with a pair of brief statements summarizing the unit's coverage and indicating what a reader can expect to derive from it in the way of practical understanding or skills. This is followed by a set of questions (**Q**) pointed more closely toward at least certain aspects of the subject at hand as a preface to systematic treatment of the topic. Hopefully the question-oriented preview will stimulate interest in what follows, offering at the same time enough of a sense of the general topic to permit a decision about whether or not it is of immediate interest.

We have attempted to keep the ensuing text as succinct as possible, addressing only the most important points and generally refraining from elaboration—avoiding, for example, consideration of pros and cons regarding a particular strategy. We frequently found it difficult to practice as much restraint as we had hoped, however, because of a keen awareness of how easily an abbreviated truth degenerates into a half-truth. Our concern with brevity sometimes forced us to limit treatment to one of several possible approaches to a situation, handling that selected as if it were—"normative". We urge the reader's indulgence in this regard, and encourage patience in the search for additional insight elsewhere (including in *The Art ...*!) when our "first aid" approach seems too superficial.

The body of the text incorporates a host of examples (**Ex**), clearly heralded by boldface marginal notations. We incidentally devoted considerable attention to layout and typographical considerations generally. An obvious case in point is extensive use of marginal "pointers", which taken together constitute a compact overview of the text as a whole. These have been deployed as an aid to the reader in search of specific information or timely examples.

Each unit culminates in a series of what we call "challenges" (**C**). In some cases these are a source of a limited amount of elaboration and/or enrichment. The conscientious reader will therefore find the challenges to be not only a test of his or her comprehension of the subject matter but also a potential source of further insight. We strongly encourage you to compare your own responses to the challenges with the set of solutions (**S**) we propose toward the end of the book.

Our criteria for selecting illustrative examples were strictly informal, but each example is designed to address in a clear-cut way the primary issue immediately at hand. In many cases the examples are products of our own collective imagination, but some were drawn from theses we have personally dealt with, or manuscripts or published papers that happen to have passed through our hands. We make no attempt to distinguish, however, between material adapted from others and free invention; moreover, we have deliberately sought to disguise the true origins of many of what are in effect "live illustrations". Thus, references are made to fictitious universities, or to "City A", "Organization B", "Compound X", or "Method Y" in situations where knowledge of the actual circumstances would contribute nothing whatsoever to the point we wish to make. Omissions are signaled by "…" or "(…)".

As stated previously, this book was conceived primarily for use by college or university students contemplating preparation of a thesis or dissertation in the natural sciences. Such documents frequently evolve into starting points for journal articles, however, so hopefully our efforts will have an impact beyond the strict confines of academic theses. We have nevertheless made no effort here to deal systematically with the peculiar demands of professional publication. We also call attention only rarely to idiosyncrasies that may affect specific disciplines.

The three of us owe a deep debt of gratitude to a host of former and current students, colleagues, and friends who generously took the time and trouble to critique preliminary versions of our manuscript, favoring us with a wealth of valuable suggestions. In this way, influences from such diverse disciplines as biology, geology, physics, and engineering have found their way into the book to help counterbalance an inherent bias toward (for us!) more familiar illustrations from our own field of chemistry.

Huntingdon, Heppenheim,   Wm. E. R.
and Schöppingen,   H. F. E.
January, 2006   C. B.

*In all our writing we prefer to avoid tiresome and repeated acknowledge-
ment—through such formulations as "he or she" or "he(she)"—of the fact
that the human race manifests itself in two distinct sexes. Surely every aspiring or committed scientist will by now be aware that women and men
play equally important roles in academia and cutting-edge research. For
this reason we regard it as pointless to persist in incorporating complicated and cumbersome modes of expression that serve only to belabor the
point. Indeed, underscoring the self-evident often ends up instead calling
it subtly into question. We hope that you, our reader, will not think ill of
us for adopting this somewhat unorthodox stance.*

Contents

Introduction

The basic idea defining a thesis in the natural sciences or engineering is rarely a brainchild of the student author. In contrast to the humanities, a science student generally chooses a topic suggested by a faculty member, who then becomes that student's "advisor". Nevertheless, as a degree candidate you retain the right to decide with whom to work, based on *your* criteria.

In choosing a topic and an advisor you are also identifying yourself with a particular subset of research initiatives, and thereby designating the corner of your discipline with which others will associate you. Choice especially of an advisor may seem an innocuous step along the academic road, but it will heavily influence the entire course of your professional development. Efforts you expend and insights you gain on the way to a successful thesis—especially a *dissertation*, the standard prerequisite for a doctorate—will make you into a specialist with a set of highly developed skills, capable of moving about rather freely in some more or less clearly defined technical domain. Your mentor will play a profound role in this once-in-a-lifetime educational experience.

You will have your own reasons for selecting a particular advisor, and for choosing among alternative thesis topics. The latter choice will probably depend on many factors beyond simply enthusiasm concerning a specific problem. Thus, you will find yourself trying to weigh the impact of such considerations as opportunities for financial support, breadth of experience, and current and future job prospects. Other relevant questions include: Do I want to be part of a large or a small research group? How closely do I wish to be supervised? What sorts of challenges have excited me in the past, and why? Where do I see my strengths lying? Do I have the confidence and stamina necessary to undertake an exceptionally speculative—or even risky—quest, where success could make me famous? It is legitimate also to look into how long a project in a particular research group typically takes to complete, and the morale and overall level of satisfaction of current group members. Incidentally: a good advisor should be anxious to suggest a number of possible problems, and even discourage you from seizing upon the first one you encounter.

Don't take the selection of either an advisor or a research topic lightly: too much is at stake! Learn about recent experiences of your peers, and examine the background of any theme you are seriously considering, scouring the library for relevant new publications. Think also about the framework within which that material would need to be interpreted. And what is the present state of related projects?

In making your choice, take into account not only the scientific "appeal" of a concept, but also how the work might dovetail with your career aspirations. Be especially wary of a project that seems only vaguely defined, with no clear perimeter. A thesis or dissertation topic—like any piece of true research—should obviously be judged to some extent on its inherent interest, but equally important is how strongly it excites your curiosity, and urgency you perceive regarding the outcome. The right problem will feel like much more than a pedestrian assignment, suitable for whoever happens along.

Ideas characterizing theses or other research publications are seldom expressed in question form, but they always rest on a foundation of inquiry. Consider a title like "Silylated Hydroxylamines: Preparation and Properties". Certainly no "question" is obvious from that label, but it masks a great many potential issues: How might silylated hydroxylamines be synthesized? Is a direct approach feasible? What properties might such compounds have? Are there inherent limits to their stability? If you were to choose this as a thesis topic, you would first need to find out everything the professional literature has to say on the subject. Then comes identifying and refining the most promising approaches to specific questions you wish to pose. Finally, the real challenge is in finding reliable answers to those questions.

As work progresses, never allow yourself to be distracted so that you lose sight of the primary goal, but at the same time stay keenly alert for unexpected insights. Indeed, both you and your advisor should—at least in the first months—maintain a sincere openness to redefining the target in light of fresh developments: unanticipated roadblocks, for example, or new revelations in the literature, but also unforeseen vistas. Your ongoing relationship with your advisor will be absolutely crucial, and you dare not knowingly jeopardize it. In particular, major decisions concerning your work should always reflect consensus.

… With these preliminary thoughts on the table, we wish you the very best in all that lies ahead, as we turn our attention to how you should go about crafting the document that will constitute the public unveiling of your success. Before getting down to serious business, however, we encourage you to savor the timely parable reproduced on page 4, borrowed (with the author's permission) from an article by E. A. Mason in the *Journal of Membrane Science* (1991) 60:125–145.

Part I
Style and Methods

A rabbit is sitting outside its hole writing its dissertation when a fox comes along.

"What are you writing?" the fox asked slyly. (Foxes are always sly.)

"My dissertation," answers the rabbit.

"And what are you writing it on?" asks the fox (slyly).

"How to eat foxes."

The fox thinks this is incredibly funny and starts to laugh, but the rabbit urges the fox to join her in her hole. The fox agrees; we see the fox slyly follow the rabbit down, and later the rabbit returns to her dissertation.

Along comes a wolf, who asks the rabbit, "What are you writing there?"

"My dissertation."

"And what are you writing it on?"

"How to eat a wolf."

"And how would a rabbit know anything about eating wolves?"

"Well," says our rabbit, "Why don't you come down into my lab and see?"

So the wolf follows the rabbit into the hole, and soon the rabbit returns to the surface alone again.

Now, if we had been able to see inside the rabbit's hole, we would have seen piles of bones and a *lion*. The moral of the story is: Don't ask, "What's your thesis on?", ask, "Who's your advisor?"

1 The Laboratory Notebook

● This unit provides information about what a laboratory notebook is and how it should be managed.

■ Adopting at the very outset of your research the suggestions offered here will help you plan your research activities, carry out the appropriate experiments, and document your efforts in a rational and effective way, taking into account both your own interests and the legitimate interests of others.

Q 1-1 How should a laboratory notebook be structured?

Q 1-2 What do we mean by "an experiment"?

Q 1-3 What considerations are important in maintaining a notebook?

Q 1-4 For whom does one prepare a notebook, and what purposes will it serve?

Q 1-5 What form should notebook entries take?

1.3, 1.4   *)

Nature, purpose   A *laboratory notebook* supplies the official documentation for all of a scientist's workaday activities—including observation, experimentation, analysis, and the like. This documentary record must be developed in a continuous, ongoing fashion, and is to include all the detail necessary to permit verification, evaluation, and further extension of the scientist's findings. The notebook thus becomes a repository of real-time, authentic records of every scientific study a researcher undertakes. These records might in turn be regarded as equivalent to "germ cells", expected later to produce reports of various kinds (see below) as precursors in most cases to publications.

Field records   In certain disciplines (e.g., archeology, ornithology) the scientific notebook more closely resembles a professional *diary*, where again the researcher

---

* The symbol at the start of this line (▷) is intended here and elsewhere to direct the reader's attention to relevant sections in *The Art of Scientific Writing* (2nd ed., 2004).

faithfully transcribes relevant observations and attendant thoughts—as these occur—but now in the context of a project based in the field as opposed to a laboratory.

Reports

For our purposes, a *report* is to be understood as a document supplying comprehensive information about some discrete investigation or research module. The *thesis* (or *dissertation*)—an indispensable prerequisite to many advanced degrees—is an especially exhaustive form of report, consisting of a formal, detailed account of a major piece of independent research. Even the laboratory notebook might be interpreted as a type of report, at least in the broadest sense, although its primary purpose is to furnish source material for *other* reports, starting with interim progress reports. The latter often supply much of the text for the crucial "Experimental" section (see Unit 14) of a subsequently prepared thesis or journal article.

The deficient notebook

If you do a poor job of maintaining your notebook you will later find it almost impossible to compile meaningful and reliable descriptions of your work. Disorganized, unclear, or illegible notebook entries at best lend themselves to laborious and unreliable interpretation, and may well compel the researcher to allocate precious time to awkward attempts at duplicating previous findings simply because the existing written record provides too little information for earlier work to be reproduced. A case in point would be the notebook account that ignores parameters indispensable to the success of a newly discovered chemical reaction, which would in turn cast serious doubt on the validity of the alleged outcome.

Getting started

The suggestions that follow are meant to help you document your scientific activities as efficiently as possible, at the same time avoiding unpleasant consequences frequently attributable to flawed records. Ideally you are encountering this advice early in your research, or were long ago the beneficiary of similar counsel from some other reliable source. Whether practiced in a laboratory, a greenhouse, a remote forest, or on a wave-tossed research vessel somewhere in the Baltic Sea, your record-keeping skills will be put to their first crucial test when the time arrives to submit an initial *interim report* (cf. also Unit 3). Any problems you encounter at that stage should be thoroughly and promptly analyzed as a precursor to immediately mending your ways.

Learning to maintain a good laboratory notebook is actually valuable preparation for other writing responsibilities to be faced further down the road. In a sense, on that fateful day when you anxiously commit yourself to the topic that is to define your graduate research, and then inscribe the first words in your notebook, you are formally embarking upon a long journey expected to culminate in a first-rate thesis or dissertation: a formidable document offering eloquent testimony to your proud fulfillment of one of the most important requirements for the degree to which you aspire.

Basic requirements One generally takes it for granted that laboratory notebooks will be *handwritten*, in part because of the importance attached to assuring that all entries are made promptly and "on the scene" (although some forward-looking laboratories are experimenting with the use of sophisticated electronic alternatives). It is essential, however, that anything you write in your notebook be clearly legible and intelligible, not only to you but to others who may have a legitimate interest in your work, thoughts we will further pursue shortly. As you go about carrying out your experiments be sure to describe every step so meticulously that all your results are reproducible on the basis of the original notebook record alone, and the ultimate source of every piece of data or observation is made crystal clear.

To whom does a notebook belong? It is a good idea to inquire at the outset whether your advisor will expect you to leave your notebook(s) behind when your tenure in the research group ends. If so, you will almost certainly want to prepare duplicates of each page as you go along, for your own future reference. As you might imagine, your advisor almost by definition has a strong vested interest in discoveries originating in his or her group, and thus every right to expect immediate, long-term access to all the associated details. Moreover, outsiders who may have a financial stake in the work, or others who contribute to it in some significant way, could also have a valid claim on both the results themselves and the documentation supporting them, claims ordinarily exercised through the leader of the group.

Timely data entry Notebook entries are relied upon to be faithful descriptions of original observations, prepared immediately after the fact. Never entrust your (important!) data to scratch paper, even temporarily, or permit yourself to become dependent on blackboard notes or—worse yet—your memory. As soon as you establish a numerical value from an analytical instrument, for example, enter it immediately into your notebook: *now*, while it is fresh in your mind. A second reading taken later, or under a different set of measurement conditions, might conceivably alter your *faith* in an observed piece of data, but it cannot possibly affect the original observation itself, which you are committed to preserving for posterity.

A scientist's diary Treat your research notebook with the respect a *diary* deserves, not a scratchpad. See that your name is prominently displayed on the cover, and scrupulously date both the beginning and the end of each entry. Over time you are likely to fill several such notebooks. If so, they should be clearly labeled with numbers establishing their proper sequence.

"Documentation" It could be that your results will someday be relevant with respect to a patent application, or otherwise serve to undergird a claim of priority. Or perhaps evidence will later be needed regarding precisely how you conducted one of your experiments. Situations such as these explain why notebook entries must necessarily be truly "documentary" in character. Notebook records have on occasion even been cited in courts of law when serious questions arose about the origin and/or validity of specific scientific information. It is thus imperative that the following guidelines always be strictly observed:

- An "official" notebook must be a *bound* volume, protected by a sturdy cover;
- Pages in a notebook must be *sequentially numbered*;
- The written record must be *continuous*, without unexplained gaps, although each individual experiment should commence on *its own new page*;
- You may as necessary *strike through* mistakes you make, but only in such a way that the original words or numbers remain legible (erasure is never tolerated);
- Unavoidable *blank spaces* should always be "cancelled" with a bold "**X**";
- Everything should be entered in *ballpoint pen*.

Physical form Under no circumstances is a ring binder or spiral-bound notebook acceptable for documentation purposes, since there is no way rigorously to exclude the possibility that pages have been removed, exchanged, or introduced after the fact. A scientist's notebook must also have a cover sufficiently robust to shield the contents from accidental damage.

It is customary for successive notebooks to be characterized by number (as I, II, III, etc.), which in turn facilitates creating unambiguous numerical identifiers for individual notebook entries. Thus, the notation

Ex 1–1 a II-26

is readily interpreted as referring to an account commencing on page 26 of notebook II. Appending a set of initials, e.g.,

b JCB II-26

makes the reference even more meaningful by distinguishing it from work of colleagues and other members of one's research group.

The notion A typical *experiment* (e.g., a chemical reaction, say, or careful observa-
of an "experiment" tion from a distance, on a particular day, of the mating behavior of a spe-

cific pair of eagles) is so designed as to elicit from nature the answers to one or more clearly defined questions. The official description of such an experiment—in the form of a unique notebook entry—must 1) in a succinct way declare the intent of the inquiry, 2) recount in detail the complete set of individual steps or determinations undertaken, and 3) reveal as unambiguously as possible whatever direct answer or answers emerged. One might almost characterize a model entry as a reproduction in miniature of the experience itself.

Structure; headings It follows from this that a typical laboratory-notebook description consists of three distinct "parts"

– a problem statement/introduction,
– a procedural narrative, and
– an account of findings

although these are not always separately labeled, or even cleanly differentiated. Each entry commences with a title or heading. For example, in the case of a chemistry project one might find pages that begin:

Ex 1–2
Synthesis (Rearrangement, Isomerization, ...) of ...
Characterization of ...
Treatment of ... with ...

An important aspect of the title is the contribution it makes to a meaningful table of contents for the notebook, although references to a particular experiment in other contexts are likely to be by page number.

This concise verbal identification should be accompanied by the applicable date (e.g., 10/14/04), possibly elaborated as

Ex 1–3
Begun 10/14/04, completed 10/16/04

Next should come a carefully crafted introductory sentence articulating the goal of the experiment

Goal, starting point In the first major "part" of a notebook entry (purpose, or introduction) the challenge is to express as succinctly and precisely as possible what the experiment in question is all about, and to summarize resources upon which it depends. In the case of a chemical transformation it is common practice to begin with a structural equation:

Ex 1–4

$$RPCl_2 \xrightarrow[?]{Li\,P(SiMe_3)Ar} R{-}P{=}P{-}Ar$$
(JcB II-24)

This bit of widely accepted shorthand, suitably annotated of course to identify "R" and "Ar", expresses in the highly condensed formula language of a chemist the message: "A dichlorophosphane is to be treated with a particular lithium aryl(trimethylsilyl) phosphide in a process expected to lead to a diphosphene." The question mark under the reaction arrow here may signify that the researcher was uncertain at the outset about whether a pair of phosphorus atoms would indeed end up joined through a double bond in the manner shown; alternatively, it could signal initial ambiguity regarding the depicted *trans* (or *anti*) stereochemistry.

Elaboration The same notebook entry might instead have been introduced by a purely verbal passage based on the "translation" presented in the preceding paragraph, starting perhaps with a phrase like

Ex 1–5 *The goal was to find out if* …

but the symbolic alternative has the obvious advantage of being much more concise. Dwelling at length on the essence of the proposed chemical transformation itself—outlining a possible mechanism, for example—is inappropriate in a notebook context. One would thus refrain from explaining here that a reaction of the type envisioned is thought to proceed by way of an intermediate, which is transformed only slowly into the desired diphosphene in the course of warming (although the *fact* that extra heat is applied will of course be disclosed at the proper point in the subsequent account!). Similarly, that this synthesis is being conducted as a "one-pot process" without isolation of the intermediate may represent an interesting choice on the researcher's part, but the rationale behind it is probably irrelevant from a laboratory-record perspective (unless testing the one-pot approach was the chief motivation for the experiment).

In the example above, the source of the indicated starting material (the dichlorophosphane $RPCl_2$) has apparently been disclosed in a previous notebook entry, as indicated by the notation "JCB II-24" below the equation.

The rest of the introductory section of a notebook entry can be used to convey sundry bits of information, especially factors that could in some way influence the experiment; e.g.,

– sources of key compounds and equipment, perhaps expanded to include details (addresses, phone numbers);
– critical specifications, or standards of purity in the case of materials;
– the precise location and timing of the work if this might somehow be significant; and
– references to background literature

as illustrated in Ex 1–6.

Ex 1–6 *Trimethylsilyl chloride (Down Chemicals)*
Carrageenan (1% w/w, Marine Colloids, Inc.) USP Grade
OSM₂-Meter (Logo Instruments)
Wistar rats (m, SPF, HC/CFHB, 212-267g, age 6-8 weeks)
Fraction 2 from JCB II-24
Collection point III, 8:00 AM, cloudy
Ref.: Winter & Porter (modified as in I-37)

The introduction is also a convenient place for such miscellany as calibration factors, sketches of one's experimental setup, or a standard protocol for dilutions.

Literature reference   The note alluding to the professional literature in Ex 1–6 ("Winter & Porter") might usefully have been expanded into a more explicit link to the relevant bibliographic data (see Unit 2).

Experimental description   What follows should be a comprehensive, personal narrative account of the experiment itself, constituting a "Part II" with respect to the overall entry. You may wish to call attention to the transition entailed by separating the two "parts" by a horizontal line.

Writing style   In order clearly to establish the "report" character of the information presented it is helpful to set descriptive passages in the past tense, and to employ the passive voice ("… such and such *was done* …"), at least to the extent that you elect to express your thoughts in complete sentences. There is no objection, incidentally, to taking advantage of familiar laboratory jargon along the way (cf. Ex 1–7).

Ex 1–7   *… the solution 1 was added …*
… was stirred over the noon hour (hood) at …
… introduced into the solution with a microliter pipette …
… heated for 5 hr and then centrifuged at 1850g for 5 min …
volatile solvents were removed w/ a rotovap, after which …

Some people prefer to write notebook entries in the first person ("I heated …"), despite the fact that this is not customary in formal reports. Feel free in any case to introduce personal reminders or admonitions to yourself (cf. Ex 1–8).

Ex 1–8   *[Talk with Frank!]*
Repeat, preferably tomorrow morning.
(New valve ordered?)
[Try a lower temperature next time]

"Unimportant"
comments

While you may elect to rely heavily on abbreviations and keywords, all descriptions must nevertheless be rich in detail. Be generous in your inclusion of informative comments like "refluxed *over the noon hour*", or notations about things that might be taken for granted by a skilled practitioner of the art (e.g., "Solution evaporated *in a hood*"). The most important principle to bear in mind is that a person reading your account must be put in a position to carry out the experiment exactly as you did.

Practical tips and tricks constitute valuable additions (cf. Ex 1–9).

Ex 1–9

The vent was opened slowly and carefully prior to ...
... rinsed from the cold finger with acetone.
... scrubbed with a long-handled brush.
... steadied with the other hand for as long as necessary.

Be sure also to include incidental observations like

Ex 1–10

... turned momentarily red ...
... felt warm to the touch ...

Technicalities,
observations

Many of these details will undoubtedly never make their way into the professional literature, and some will not even appear in your thesis, but you should nonetheless err on the side of too much information in your laboratory notebook. Accounts you prepare later for yourself and others can always be condensed or "massaged" as may seem appropriate, but the original record must stand as it is, without elaboration. One often becomes aware only in retrospect that some act or observation which seemed almost trivial at the time is in fact critical—and far from self-evident. A quest for such (unpublished) technical details is the most common motivating factor behind someone's delving into a half-forgotten stack of laboratory notebooks, and the chief reason why notebooks tend to be carefully preserved over the course of many years.

Findings, results

Accounts in your notebook should make clear not only what you *did* in attempts to answer the questions at hand, but also what answers you found. These constitute "Part III" of the entry: the "results". With a chemical synthesis, for example, results in this sense might consist largely of the weight of isolated product, a melting point, and spectral data you acquired, whereas other types of work lead primarily to qualitative, descriptive answers. As with the introduction, the results section may or may not be specifically singled out as a discrete element separate from the experimental details.

Measured values

Experimental results in many cases come in the guise of numerical data. Here again, the watchword must be fidelity: a digital readout or pointer position should be reproduced just as you encounter it (with an analytical instrument, perhaps), in precisely the form in which it was presented to you. Converted data or derived quantities are necessarily the product of one or more subsequent processing step, which must always be documented separately.

With an investigation characterized by a great many routine measurements, it may suffice to describe in detail one particular determination, in a single place in the notebook, thereby avoiding tedious repetition throughout the course of an investigation—which could extend over a long period of time. References to this master protocol can then serve as "place-holders" for suitable elaboration later as necessary in some other context (i.e., a report or publication), as illustrated for an interim report in Ex 1–11.

Ex 1–11

The concentrations of pollutants a, b, c, and d were monitored continuously in Little Salmon Creek throughout the summer of 2003 (June to September) at sampling stations 1, 2, 3, and 4 with the existing on-site autoanalyzers (see JcB I-9 and I-10). Additional determinations were made of pesticides e and f at the headwaters (1) and mouth (4) of the stream. Results are shown in plots JcB I-15-1 through JcB I-15-4. Daily maximum values are summarized (along with relevant times of day) in Table 12.

Statistics   In the event that you collect multiple values for a given variable, be sure to note each one individually, and only *then* compute an average, even if an average is all you intend ultimately to report publicly, perhaps in conjunction with a standard deviation and a tally of the contributing values.

Along the lines of preparing a faithful record of all accumulated data—in the form originally acquired—note that a weighing, for example, is ordinarily the product of *two* pieces of data (gross and tare values), from which a net value is then *calculated*. All three numbers should therefore appear in your notebook (Ex 1–12).

Ex 1–12

Gross   25.37143 g
Tare   25.29781 g
_____
Net   73.62 mg

Similarly, for a manual titration involving a burette you should document not simply the amount of titrant added, but the actual meniscus levels observed initially and upon reaching the equivalence point. (This example may seem archaic in the context of a modern laboratory, but it vividly illustrates the point.)

Data workup Quantities whose values are ascertained through some sort of (manual) data manipulation or conversion process (the area under a peak from a chromatogram, for example) should be accompanied by careful explanation of all the steps entailed, together with any numerical factors utilized, permitting this aspect of your work to be accurately reproduced as well. The "results" section of a notebook entry is also the logical place to document any statistical analyses you may undertake in the interest of establishing a quantitative measure of reproducibility, for example, perhaps as part of a test of significance.

Instrument error With data derived from analytical instruments or similar devices it is important always to be explicit about calibration factors you may have utilized. If you find yourself working with one of several similar instruments, be sure you record *which* one, by specifying a location or some other identifying characteristic. Precautions like these are often the key to accounting eventually for perplexing "outliers" in a data set.

Future plans The notebook description of an experiment may conclude with a few observations pointing toward the future. For example, the researcher responsible for the synthesis outlined in Ex 1–4 might wish to append a note regarding spectral analyses still to be undertaken for establishing whether the product isolated in fact conforms to the assumed *E* configuration, or if both substituents are instead located on the same side of the phosphorus–phosphorus double bond (in a *Z*-arrangement). Or possibly a reminder is in order that an analogous reaction should be carried out as soon as possible with the corresponding di*bromo*phosphane as starting material (as experiment JCB II-27, perhaps).

Interim reports, experimental sections Devoting a certain amount of attention to *form* in the preparation of one's notebook write-ups can pay handsome dividends during subsequent development of the "Experimental" portion of an interim report or a thesis. Ideally, little more will be required than elimination of excess baggage, a bit of reorganization, and some rephrasing. Free-standing keywords and sentence fragments in a notebook would of course need to be transformed into complete sentences, and preliminary results can often be upgraded to convey greater meaning (e.g., conversion of a yield in milligrams into a value expressed in moles, to permit interpretation as a percentage of a theoretical yield). A typical reworking of a notebook account is illustrated in Ex 1–13, with part **a** representing a notebook excerpt and part **b** the corresponding passage from a formal report.

Ex 1–13 a   – 56 –

8/25/03

Oxidation of Ketone 6a (con'd from pg. 53)
Reaction mixture filtered (Büchner funnel)
crystallized
Solid (pale yellow), 1.63 g

Solubility Tests

Acetone	*insoluble*
Et_2O	*insoluble*
CCl_4	*insoluble*
H_2O	*soluble (esp. hot)*
CH_3OH	*very soluble*
	(ca. 300 mg in 2 ml)

Recrystallized from 30 ml CH_3OH/H_2O
                (ca. 1:3), colorless needles

8/26/03

Dried 24 h over P_4O_{10} → 1.40 g (86%)
m.p. 72–76 °C (crystals collapse at 65°C)
no decomposition to 150 °C
Tests acid in aq. soln. (pH paper)
> 8.5 mg in 50 ml H_2O (doubly distilled):
        pH 4.65 (pH meter #3)

b ... Crude product (pale yellow) was isolated from the reaction mixture in crystal-line form. An analytical sample of **3a** was obtained by recrystallization from a water–methanol mixture ($\varphi = 25\%$) as colorless needles (m.p. 72–76 °C; Jones 1980: 78 °C) in a yield of 75.8%. Aqueous solutions of the recrystallized material are distinctly acidic ($pK_a = 9$).

The text for the report version obviously reflects considerable "polish" and formality relative to its notebook counterpart, although both are characterized by a high level of information density. The writing style adopted for this report is consistent with that generally found in the "Experimental Section" of a journal article, and would be entirely appropriate for a dissertation as well.

"Documentation" The end of every notebook entry should be signaled by a bold horizontal line, with equally prominent diagonal lines (or "X"-like markings) used to cancel whatever blank space remains at the bottom of the page. These are measures that once again reflect the "documentary" character of the content, declaring in effect that each experiment should be regarded as "complete as described." The record applicable to the next experiment—which in the case of a chemical synthesis might describe isolation and identification of a byproduct—should be consigned to a fresh page, which must in turn be properly dated.

Wider ramifications

A single experimental write-up ordinarily should not exceed one or two notebook pages, preferably a left and right page *pair* for convenience. If a project threatens to consume more space than this, try subdividing it into two or three pieces, each to be treated as an experiment in its own right.

The optimal length of an entry

As already suggested, one's laboratory notebook—properly exploited— can evolve into a valuable organizational tool with respect to all of one's laboratory-related activities. For instance, samples or other artifacts derived from a particular experiment are usefully catalogued and labeled in terms of the notebook page where they are first mentioned (e.g., "JCB II-26"), and the same applies to electronically generated records like spectra, chromatograms, or computer printouts. Multiple items from a single experiment are easily distinguished by building upon the basic identifier, as illustrated in Ex 1–14.

Ex 1–14

JcB II-26-1
JcB I-47-IR-A

The labels above would thus point to the first sample preserved from an experiment commencing on page 26 of notebook JCB II and an infrared (IR) spectrum defined on page 47 in notebook I. Labeling of this sort can contribute in an important way to establishing and preserving order. Your laboratory notebook—or project journal, or field record, or whatever you choose to call it—thereby develops into a centralized management tool responsible for a sense of discipline throughout your research domain. It seems especially appropriate to underscore this aspect of the research notebook at a time when computers have come to shoulder much of the burden formerly borne by the practicing scientist, largely eliminating manual effort that once was everywhere taken for granted. Computerization makes it all the more important that you devise a standardized system for keeping track of all that transpires in the course of your work. What could be better suited to giving shape to this system than the scientist's notebook—which has always been revered as the researcher's professional diary, preserving and coordinating all the salient facts.

◇

C 1–1

Why are the pages of a laboratory notebook numbered? What benefits does page numbering confer?

C 1–2   Identify steps you should take in the context of maximizing the "documentary" character of your notebook.

C 1–3   Why are blank portions of notebook pages always "crossed out"?

C 1–4   What purpose is served by careful dating of one's notebook records?

C 1–5   Why would it be inappropriate to recruit a ring binder to serve as a laboratory notebook?

C 1–6   Is it always clear where the boundaries should lie between one experiment and the next?

C 1–7   Ideally, how much notebook space should a single entry be allowed to claim?

C 1–8   Identify some of the most important principles to bear in mind as you formulate the notebook account of an experiment.

C 1–9   To what extent is "laboratory jargon" acceptable in a notebook? Cite a few examples of common jargon from your particular discipline.

C 1–10   Why should each notebook entry be assigned a title?

C 1–11   Identify three fundamental criteria to be fulfilled by the notebook description of every experiment.

C 1–12   Of what structural components ("parts") does a notebook entry typically consist?

C 1–13   Provide examples of information that would logically fall within the "first part" of a notebook description for an experiment in your discipline.

C 1–14   Why in the future might someone be interested in reading entries from your laboratory notebook?

C 1–15   Compare the sample notebook text in Ex 1–13 with the accompanying excerpt from a report. What are some of the most striking differences?

2 Literature Work

● Here we consider how you should begin dealing with the professional literature in preparation for embarking upon thesis research.

■ Mastery of the material discussed will assist you in locating all the relevant information already available in published form and then organizing it so you can apply it directly to your own work, developing in the process a personal collection of literature sources.

Q 2–1 What does a scientist understand by the term "the literature"?

Q 2–2 Why is it important to be intimately familiar with the structure of the literature of your discipline and with its most important components?

Q 2–3 What role must the literature come to play with respect to your research and the thesis you plan ultimately to write?

Q 2–4 How should you go about acquiring an overview of the various publications with a direct bearing on your efforts?

Q 2–5 How can customized computer-based resources contribute to personal literature management?

 9.1, 9.2

"The Literature" Almost by definition, advances in science come as the outgrowth of discoveries made earlier. The sum of all publicly-available knowledge within a given discipline — i.e., *published* information—is what constitutes that discipline's *literature*. Every field, indeed even every subfield, has in this sense a "literature" of its own. In the course of your studies you will already have encountered portions of the literature, making direct use of it as you worked your way through textbooks, or repeated experiments described for you in laboratory manuals.

Familiarization with the literature The very first challenge you must confront in thesis research is becoming intimately familiar with all the existing publications related closely to your chosen topic. Getting a good start on that process is absolutely prerequisite to conducting research effectively. It is essential that you be fully apprised of what other investigators have already learned about your subject (and how they learned it!), and only with this background will you later be able to fit your own results properly into the larger matrix that constitutes reality (see Unit 12). The more you have availed yourself during your studies of opportunities to become acquainted with key handbooks, reference works, and journals in your field, the easier it will be for you to meet the literature hurdle you now face. For openers, you must of course discover where and how someone in your position goes about accessing the various resources.

Getting into your topic | Generally speaking, students in the natural sciences and engineering actually spend considerably less time in the library than their counterparts in the humanities, but in the beginning you will still need to devote at least several intense days exclusively to "literature orientation". Your advisor will undoubtedly have pointed you toward a few of the most indispensable sources, typically articles that have appeared in various research journals. While these may seem a tempting place to start, we encourage you instead to step back a bit and consider your topic from a broader vantage point— by consulting a few fundamental textbooks, perhaps, as well as relevant entries in scientific or even general-purpose encyclopedias. Incidentally: once you agree to pursue a topic proposed to you by a faculty member you automatically become a member of that person's "research group". Right away you should check to see what sources of information, if any, membership in that group may put at your disposal. Of particular interest will probably be theses and dissertations prepared by your predecessors, especially since you are likely to find yourself building upon some of their results.

The overview | Discuss with your advisor the extent to which you should immerse yourself in the professional literature prior to planning and conducting your first actual experiments. Ultimately, you alone will be responsible for diligently unearthing and interpreting every scrap of published information with a significant bearing on your topic. With luck (coupled with an effective search strategy) you might manage early on to locate a timely and highly relevant "review article" prepared by some expert in the field. This would have the great advantage of providing you immediately with a substantial collection of valuable references, especially published reports from those responsible for establishing the salient facts in the first place ("primary sources"). Don't be hesitant to seek help (from library personnel as well as your coworkers) regarding the most promising places to look for such reviews in your particular discipline.

If you find that your research group is not in a position to offer significant literature support, and you can't locate a suitable review, then your only option is to begin assembling from scratch your own comprehensive overview. In a worst-case scenario you might during the process discover that the topic you selected has already been explored more extensively than you (and your advisor!) realized, or perhaps you will run across ominous hints pointing to unforeseen major obstacles. Either eventuality would call for prompt and frank discussion of the situation with your advisor—after of course carefully reexamining the validity of your concerns to minimize the risk of embarrassment.

The heart of the scientific literature is the *journal*, the medium virtually every scientist relies upon for wide dissemination of his or her research results. You should be forewarned that reading and fully comprehending the content of a journal article is, especially for the beginner, a complicated, arduous, and often very frustrating exercise, but you must force yourself to persevere, because all too often "the devil is in the (initially obscure) details", and you are obliged to master them if you hope to succeed as a researcher.

The most highly respected journals operate under the principle of "peer review", which means articles are published only if they meet specific quality criteria defined by the editor(s) and monitored by other scientists the editor regards as qualified to pass judgement on manuscripts submitted for possible publication. Unfortunately for someone in your position, the list of journals potentially containing relevant information is likely to be vast, so searching the various publications directly for useful material is highly impractical.

Probably the most important *indirect* entry points into the scientific literature are several exhaustive "abstracting journals", including *Biological Abstracts*, *Chemical Abstracts*, and *Physics Abstracts*, which at frequent intervals issue brief summaries of virtually everything published within the general subject area they represent. One locates abstracts of relevant documents with the aid of exhaustive subject indexes, in conjunction with an arsenal of carefully selected, definitive keywords or technical terms. Until relatively recently, the only way to deal with the abstract literature was in printed form, a single year's worth of which (including indexes) could consume over a dozen feet of library shelf space. Now almost every researcher has access to the abstracts through online databases, resulting in countless hours saved. An example is the "SciFinder Scholar" service devised for *Chemical Abstracts* and made available by the American

Chemical Society. As you might anticipate, such databases are not open to unlimited access by the general public, but operate instead on a subscription basis. As a student you will in all probability be able to utilize these facilities with the aid of subscriptions maintained by your university library. Library staff will be more than happy to assist you.

SCI   Another valuable window on the literature, one with which you are less likely already to be familiar, is *Science Citation Index* (*SCI*), a product of Thomson–ISI. This resource, like the abstracts, is generally available through university libraries and accessed (also like the abstracts) via the Internet, either with the help of a wide-ranging database-provider like Dialog (another Thomson enterprise) or STN International, or as one component of the Thomson–ISI *Web of Science* or *Web of Knowledge* services.

In essence, *SCI* is an "index of footnotes" (more accurately, *citations*, which commonly appear at the *end* of a journal article). Every citation printed in every article from virtually every important scientific journal in the world is duly recorded by the editors of *SCI*, together with all the bibliographic data required to locate the respective "source" articles.

Working with *SCI*   Let us assume you are already acquainted with at least one journal article closely related to your research interests—even a relatively old one. With this starting point *SCI* may prove to be a godsend, since it can point you in the direction of all those papers of more recent origin (!) that have referenced (i.e., *cited*) the paper of interest, many of which are certain to be of great interest to you as well. "Citation searching" of this sort often leads to valuable "finds" one could never hope to unearth in the course of a more conventional search. An *SCI* search can obviously be "targeted" more precisely by singling out papers that have simultaneously referenced two or more of your most important resources.

Keyword searches   Alternatively, one can search within the *SCI* files for articles associated (through their titles, for example) with promising keywords, although an equivalent set of results can be compiled just as easily on the basis of the abstract literature.

Conference Proceedings   Quite apart from journal articles, there is a second type of "primary publication" you should also try to monitor: information presented orally in the context of major professional meetings or specialist's conferences. Tapping this source has in recent years (since 1982) become more practical thanks to a special compilation known originally as the *Index to Scientific and Technical Proceedings* and now included in the ISI *Web of Knowledge* working environment under the rather more generic heading *Proceedings*.

Personal literature files

Even if you find that a fairly extensive collection of literature resources is available to you in communal form within your research group you will undoubtedly still want to create and maintain a personal literature file of your own. This can in principle be accomplished in a variety of ways. We begin by describing at some length a proven, traditional approach based on ordinary file cards, but the underlying strategy lends itself admirably to computer adaptation as well. Thus, if you reject the prospect of working with handwritten cards as hopelessly archaic, you should opt from the beginning to assemble your literature database on a PC.

The key to the entire system is an *author file*. You must thus begin by

– creating for every relevant research publication you read or intend to read—journal articles, books, conference papers, etc.—a dedicated reference "file card" (database record): one publication, one card (record). Enter into that record the name(s) of the corresponding author(s), together with all the basic information required to pinpoint (unambiguously!) the document in question. These "bibliographic data" are precisely what you will need later in order correctly to cite that particular source in the context of your thesis (see Unit 15).
– Assign to each record as you create it a sequential "record number".
– In the case of records in card form, arrange the stack alphabetically by author (or, as necessary, first author/second author) in a sturdy file box, treating the year of publication as a secondary organizational criterion (cf. the *name–date system* for citations, described in detail in Unit 15).
– Now construct a separate *numerical index* that will allow you to access your author-based file directly by record number.
– At your discretion, add to the cards provisions for whatever other information you think might be helpful: keywords, for example, or possibly even abstracts for at least some of the sources.

A model "literature card"

A sample card meeting the criteria described is shown as Fig. 2–1.

Literature notebooks

In addition to this master directory, consisting largely of bibliographic data, you will also want to develop a system for managing reprints and/or photocopies of some of your more valuable sources as these accumulate —both articles in their entirety and appropriate excerpts. We suggest the following measures in this context:

– Each such item you propose to add to your collection should be annotated with the most critical bibliographic data (perhaps in abbreviated form) together with the record number assigned to that document in your card file (or database).

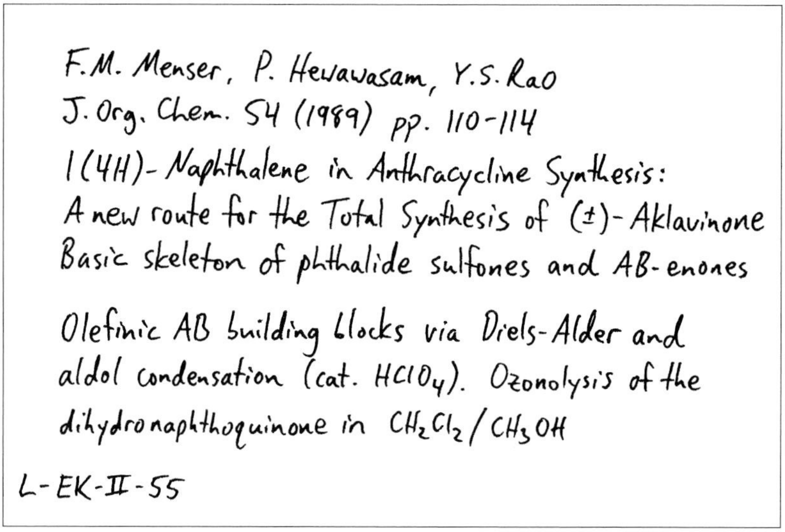

Fig. 2–1. Example of a literature card bearing a unique record number (top right) as well as a link to a collection of reprints (bottom left).

The card reads:

> F.M. Menser, P. Hewawasam, Y.S. Rao
> J. Org. Chem. 54 (1989) pp. 110–114
> 1(4H)-Naphthalene in Anthracycline Synthesis:
> A new route for the Total Synthesis of (±)-Aklavinone
> Basic skeleton of phthalide sulfones and AB-enones
>
> Olefinic AB building blocks via Diels-Alder and aldol condensation (cat. HClO₄). Ozonolysis of the dihydronaphthoquinone in CH₂Cl₂/CH₃OH
>
> L-EK-II-55

- Gather up the various materials in sturdy notebooks (ring binders are ideal), numbering the pages consecutively.
- Annotate the corresponding master file-card/database record for each such document with the applicable notebook-page location.
- If you later wish to add additional pages related to an already stored document, assign new page numbers as necessary—again to be duly noted on the corresponding file cards (and on the original documents!).

An alternative is to organize the set of literature notebooks sequentially on the basis of your database record numbers, in which case separate document-page numbers would be superfluous.

As your collection expands, be sure also to assign sequential numbers to the notebooks themselves. Extensive document collections of this sort have now become commonplace due to ever-increasing reliance on photocopiers, and as more and more information is downloaded and printed from the Internet.

Record numbers If you make a conscientious effort to follow the advice we offer you will soon find yourself in possession of an invaluable set of reference tools with a host of applications. For example, in addition to their obvious role in conjunction with your repository of bibliographic data, record numbers you assign will make it easy for you to establish cross-references among documents. Moreover, as already noted, these numbers can contribute in an important way to the organization of a collection of reprints and photocopies, and they serve as handy "placeholders" in early drafts of manuscripts you prepare—for eventual replacement by numbers keyed instead to the requisite bibliography (see Unit 15).

Computer support A literature collection based literally on file cards can of course become extremely unwieldy, but it suffers from the even more serious limitation of not lending itself well to content searches by topic or keyword. This is an arena in which a computerized database really shines. Digital technology will allow you easily to find specific records on the basis of any number of criteria, including combinations—almost instantaneously. Thus, you might decide to search all titles for the presence of a specific word or phrase, for example, or quickly isolate all the works you have catalogued that are associated with a particular author, regardless of where that person's name happens to fall in an author list. Assigned keywords also take on far greater significance with a computer-based file, whether these be selected haphazardly in the course of creating the various records (or later!), or taken from "official" keyword and sub-keyword lists developed by others: the editors of your favorite journal, for example, or one of the documentation systems considered authoritative within your discipline (e.g., *Chemical Abstracts*, *Biological Abstracts*, *Index Medicus*).

Dedicated bibliographic software Finally, we would be remiss if we failed to point out one somewhat more sophisticated approach to literature files: taking advantage of dedicated bibliographic software such as the popular program ENDNOTE (Thomson Researchsoft). Systems in this category are able not only to create and manage records like those we have been describing—in conjunction with documents of every conceivable sort (books, articles, theses, digital sources, etc.)—but can on demand also *format* stored bibliographic data automatically, in essentially any way required (cf. Unit 15), and then attend to proper incorporation into text documents of the appropriate references and citations. Software of this sort is likely to require an initial investment on your part, but you will be astonished by the dividends it can pay.

C 2–1   What is meant by the term "literature collection"? In what other ways might the same notion be expressed?

C 2–2   Why are "author files" generally selected as the basis for technical literature collections rather than "subject files"?

C 2–3   Identify some of the ways "document numbers" can prove useful.

C 2–4   List data categories you perceive as essential in the context of literature records, as well as others that, while convenient, would class as optional.

C 2–5   What is the most efficient approach to locating within a literature file all the records sharing a particular characteristic?

C 2–6   Develop a grid of properties, concepts, topics, etc., that you think would suffice for searches you might have reason to pursue in a literature file.

3 Getting Started: Outline and First Draft

- ● In this unit you will be introduced to some things best attended to in the earliest stages of writing your thesis or dissertation.

- ■ The content of the unit should provide you with a feel for the importance of preparation and planning whenever a major document is called for, but it should also help dispel some of your anxieties.

Q 3–1 When should one in fact begin the process of writing a thesis?

Q 3–2 What sorts of reference materials will you need to have handy as you work?

Q 3–3 Are there specific guidelines to be followed?

Q 3–4 How should a thesis be structured?

Q 3–5 What is one effective approach to getting your thoughts on paper?

 1.4.2, 2.3.1

Writing up one's work Most scientific professionals have rather fond recollections of the time they spent as students conducting research in the laboratory (or the field). One generally had a fairly clear vision of exactly what it was that needed to be accomplished next, and in most cases there was a reliable source of sympathetic advice and counsel nearby. On the other hand, "writing up" the results, and in the process persuasively arguing their significance, is now likely to be remembered as a long, distinctly lonely undertaking: nerve-wracking at times, and deadly dull at others. Various external pressures could be daunting, and panic threatened when carefully accumulated bits of evidence refused to coalesce to form a neat pattern.

Reliance on a laboratory notebook and a literature collection Initial preparations for writing a thesis should actually begin the day you commit yourself to your research topic, not months or years later when you think you have hung up your lab coat for the last time. Thus, you will eventually recognize that the two most important prerequisites are a set of well-maintained laboratory notebooks (see Unit 1) and a personalized literature file (Unit 2) that is both all-inclusive and up-to-date.

Interim reports   Discovering at the last minute that you overlooked—and thus never carried out—some key experiment can be devastating, as can learning that you failed to detect a crucial journal article bearing very directly on your work. The best defense against traumatic moments like these is forcing yourself at regular intervals to prepare comprehensive interim reports—on a quarterly basis, for example, in the case of doctoral research. You will thereby ensure that both you and your advisor stay fully apprised of how your work is proceeding, and at the same time effectively monitor the extent to which your gradually emerging results are interpretable within a coherent framework. The attempt to formulate an informative and convincing report sometimes leads one to an awareness that a radical change in research strategy is called for, or that a previously unexplored methodological approach has the potential to shed valuable new light on your quest—and there is also a good chance that in the course of gathering together your results you will detect a weak or missing link in the argumentation, leading you perhaps to devise and carry out an important supplementary experiment. Conscientious and timely preparation of interim reports is an obligation you should impose upon *yourself*, although your advisor may well require such reports of you in any case.

The format of an interim report   Interim reports occupy a unique literary niche halfway between the laboratory notebook and more "public" documents like theses or journal articles. Typically they are rather formal in character, and are prepared with considerable care using a word processor and high-quality letter-size paper. The text should probably be double-spaced, although a spacing of one-and-one-half lines could also be acceptable. Each page of a report should bear a prominent heading identifying the author, indicating a date of submission, and containing both a general title characterizing the overall research project and a separate report title specifying that particular document's content. A report number may be appropriate as well. As examples, consider:

Ex 3–1   Microfauna of the Juniata River
Masters Thesis Project: James Miller
3rd Interim Report, 15 October 2004
Continuing field and laboratory studies; BOD data

Ex 3–2   The Mechanism of Deoxymercuration
Cynthia Gordon
Report 1: Configuration of the 2-acetoxy-3-mercurinorbornane
  derived from norbornene
Princeton, March 2004

Reflection, and the introduction
Before you begin writing, take time in effect to "relive" and critically evaluate the various relevant notebook entries. This should help you obtain the breadth of perspective you will need if you are to compose a meaningful report. Start, by way of introduction, with a brief summary of the background and goals of the overall project, including appropriate reference to earlier reports. This is also a good place to identify important literature sources you have recently located.

Narrative
What follows should be a lucid narrative describing all your accomplishments since submitting your last report. Be sure to direct explicit attention to results you regard as especially significant, as well as to conclusions you are able to draw. Adopt a writing style essentially analogous to that in a typical journal article.

Experimental details
Finally, provide an informative synopsis of every experiment you think worthy of inclusion, with the accounts based directly on your notebook records. In the case of an experiment you carried out multiple times you will probably want to limit yourself here to one specific case you consider representative. With respect to experimental details, or interesting observations, or special circumstances that could conceivably be relevant, descriptions accompanying a report should be more inclusive than would be appropriate in a formal publication (cf. also Unit 14).

Writing style
Descriptions of this sort are generally phrased in such a way that the role of the experimenter or observer retreats far into the background; e.g.,

Ex 3–3
The compound crystallized as colorless needles melting at 72.5–73 °C without obvious decomposition.

This practice is actually as it should be, since scientific results are supposed to be objective, and *not* dependent upon the experimenter who obtains them. For the same reason, it is common to make extensive use of the passive voice; e.g.

Ex 3–4
… was determined with the aid of a densitometer.

You should also pay close attention to temporal implications in your writing. Consider, for instance, the sample text in Ex 1–12 b in Unit 1, in which we illustrate transformation of a notebook entry into text for a report. Note that in the last sentence the writer switches abruptly to the present tense, underscoring the fact that the acidic behavior of the compound in question is assumed to be an enduring characteristic, not a phenomenon uniquely associated with the day of the experiment. Changes in tense should be infrequent, however, and certainly not arbitrary.

If you were already to take some of these stylistic considerations into account in the course of composing notebook entries, less revision would be required during subsequent preparation of reports. Incidentally: experimental accounts prepared for a report can often be imported virtually unchanged into a thesis.

The optimal writing environment

No effort should be diverted toward writing the thesis itself until *all* experimental efforts are concluded. When the time for writing does arrive, however, it is important for you to have staked out a quiet, secluded place in which to work. Above all, seek to minimize the likelihood of distractions, since success will depend more than anything else upon your ability to concentrate—uninterruptedly—on this single intellectual responsibility.

Projected work schedule

Be sure also that ample time will be at your disposal; in particular, guard against the common tendency to seriously underestimate the magnitude of the task ahead. As your first concrete step, develop a tentative but realistic projected schedule. If, for example, you allocate a total of two months to the writing process, with a dissertation that would mean consistently turning out something on the order of two pages a day. Two pages might at first seem trivial, but remember that editing, revising, and polishing will be considerably more time-consuming than rendering a first draft, and you must also allow sufficient time for preparing a finished copy. Make certain, by the way, that your tentative schedule reflects a reasonable submission date, which could be a function of factors beyond your control (e.g., departmental regulations, your adviser's vacation plans, etc.). You will obviously want to avoid investing excessive time at this juncture in background reading, but assuming you have in the past taken documentation appropriately into account and conscientiously kept up with the literature, there should be little need now for prolonged sojourns in the library.

Your link to the lab

It is generally a mistake to be overly hasty about abandoning one's laboratory quarters. If, for example, you have relied upon the ready availability of some complex array of apparatus, leave it intact for the time being. Despite the best of intentions you may find out you really *need* to run some crucial experiment one more time, or perhaps you will suddenly recognize that one further piece of data could significantly bolster your arguments.

Establishing an outline

Begin the actual writing process by formulating a (tentative) general organizational scheme for the thesis or dissertation as a whole. In many cases a reasonable sequence of subsections would resemble

– Abstract
– Introduction

– Results
– Discussion
– Conclusions
– Experimental
– Bibliography

This particular pattern is in fact relatively common, and often recommended by editors of scientific journals as a model around which to build manuscripts in general. (Note the interesting implication that it can be less difficult than you might imagine to adapt a dissertation for later release as a journal article, entailing mainly a great deal of abridgement.) The various subsections reflected in your scheme should then be assigned sequential numbers (see below). Those that will encompass the largest blocks of content—typically the Results, Discussion, and Experimental sections—will at some point almost certainly become subject to further subdivision.

There is of course nothing sacred about the specific arrangement suggested above; indeed, an alternative structure might well offer significant advantages in your particular case. Unit 8 provides a more detailed consideration of thesis structure as this manifests itself in a table of contents.

The "cluster" method

How should one proceed to organize one's thoughts preparatory to writing? What is the scope of the material that actually needs to be included? In what order should the many experiments be described? (Probably *not* chronologically!) It is important that questions of this nature be thoroughly addressed before you commence serious writing, but how? Many find it helpful at this stage to take a blank sheet of paper and jot down seemingly at random—scattered across the page—keywords evoking all the various ideas that come to mind, just as they penetrate your consciousness. Others prefer to distribute such notes over a series of pages, with one page assigned to each obvious major topic. In either case you will probably discover that words conjuring up ideas with a common thread will tend spontaneously to appear in clusters—"idea clusters"—as a direct consequence of the way the corresponding thoughts are stored in your brain.

Establishing order

When you finally run out of ideas to include, look closely at your notes and begin establishing a bit more order. For example, sketch in a set of roughly circular boundaries surrounding groups of keywords that all point toward a single issue. Then add lines tying together conceptually related circles, and equip the lines with arrow points to suggest logical trains of thought. Finally, incorporate your scheme into the skeletal structure you devised previously, assigning numbers to the various elements as appropriate (see below). Soon you will discover you have before you precisely the basis you need for constructing a coherent thesis.

Throughout the foregoing process, let your imagination be stimulated and directed by questions like:

– Why should the reader take an interest in this research? (Introduction)
– Upon what intellectual foundation does the endeavor rest? (Introduction)
– Toward what concrete questions were my efforts directed? (Introduction)
– What tools were instrumental in my search for answers, and how were they deployed? (Introduction and/or Discussion; Experimental)
– What answers did I in fact uncover? (Results)
– How can my findings best be interpreted? (Discussion)
– Do my discoveries in turn raise intriguing new questions? (Discussion)
– What conclusions do I wish the reader to draw, and what fresh insights have arisen out of my work? (Conclusions)

Outlining facilities   As an alternative (or adjunct) to the cluster method you may find it helpful to exploit outlining capabilities inherent in your word-processing software. In the case of Microsoft WORD, for example, this entails first selecting the option "Outline" under the "View" menu. Especially convenient is the fact that, in outline mode, body text can instantly be directed precisely where it belongs in a developing document—and sentences or paragraphs can be shifted about with ease thanks to navigational features at your disposal.

As suggested above, order can be conferred upon your efforts and subsequently reinforced by assigning (complex) sequential numbers to the various topics and subtopics as you identify them. Thus, a number like

Ex 3–5   2.5.3

is used to designate the third *sub-sub*section under the fifth subsection in chapter 2. As your document grows and matures you will almost certainly find it necessary to modify somewhat the initially proposed structure, perhaps adding new headings here and there, or adjusting the sequence. The end result—after a certain amount of further "tweaking"—will resemble the "Table of Contents" that will eventually introduce the reader to your *magnum opus* and serve as a kind of overview (cf. Unit 8).

Starting officially   At last the time will have come to commence serious writing. Starting any
to write   earlier would almost certainly have resulted in a great deal of wasted time— a precious commodity you can ill afford to squander! It actually makes very little difference *where* in the outline you begin inserting text. The point

is that now no matter where you choose to dive in you will not be faced with the formidable barrier of a blank sheet of paper.

The first draft   Write as swiftly as you can, directing no special attention whatsoever to grammatical perfection or stylistic nuances. Elegance in this respect will be attended to later (see Units 4 and 5). For now all that matters is giving substance to your ideas—fluently, intelligibly, and in a pattern that makes sense. Bear in mind that what you are currently preparing will in any event be simply the first of many, many "drafts".

Looking ahead: It is a good idea, by the way, even as you begin, to think a bit about how final copy   the final version of your document will be produced (see Unit 5). Will you yourself type the text into a computer and apply the necessary formatting, or will someone else assist you? In the latter case, in what form will that person expect to receive your manuscript?

Institutional   Finally, by all means also take the time and trouble *now* to secure and guidelines   meticulously *inspect* a copy of all the rules and regulations applicable to theses or dissertations at your college or university—or even in your department! These vary greatly from situation to situation, and your work will be expected to conform to them in every detail—which may have serious implications even in the early stages of writing. Close examination of theses submitted by recent alumni of your research group can prove very instructive in this context.

C 3–1   List materials you will need to have at hand as you embark upon the writing process. What measures should you take to protect yourself in the case of their loss or damage?

C 3–2   At what point is it appropriate to begin actual writing? Will optimal timing of the process necessarily be entirely in your hands?

C 3–3   Imagine for a moment that your laboratory assignment has been to collect measurements on some specific internal organ as this manifests itself in several related animal species. You are also expected to take note of significant qualitative similarities and differences from specimen to specimen. In particular, you and your advisor were hoping to learn more about the function of this particular organ on the basis of its structural characteristics as these are affected by lifestyle and environment. The project required you to collect and prepare the various specimens yourself, and to carry out analyses on tissue slices obtained with a microtome. In order to take three-dimensional aspects adequately into account you found it necessary not only to obtain length, breadth, and thickness measurements, but also to devise a convenient coordinate system within which to operate. A statistical program was enlisted for evaluating the precision of your coordinate measurements, at the same time allowing you to minimize potential distraction arising out of random scatter. Another of your assignments, which also required data collection, was mapping the distribution patterns of receptor cells and neurons within the various specimens. Assuming you now have all the requisite information in hand, how might you organize the document that is to proclaim and extol your achievements?

4 Writing Style

- This unit directs attention in particular to weaknesses especially common in scientific and technical writing.

- The ideas presented should stimulate you to examine your own writing more critically in the interest of avoiding at least a few fundamental types of shortcomings.

Q 4–1 To what extent is good writing style something that can be learned?

Q 4–2 Why must one be extremely sensitive to language in the attempt to communicate important ideas?

Q 4–3 What are some linguistic issues important specifically with respect to science and technology?

Q 4–4 To what characteristics of your sentences and paragraphs should you pay special attention?

Q 4–5 What benefits can be derived from carefully considering the way a long document like a thesis is subdivided?

▷ 1.4.2, 2.3.1

Words Our goal in this unit is to approach language aspects of scientific writing, including the pitfalls, as pragmatically as we can. We thus begin with the smallest linguistic unit: the word. Words themselves are often a major source of problems, since poor word choice can foreshadow misinterpretation. The "wrong" word rarely succeeds in conveying the intended message accurately. Unfortunately, the same can be said about the *right* word— if it is used incorrectly. Finally, words sometimes are thrown into text unnecessarily, while other words languish through neglect.

Dictionaries As was pointed out in Unit 3, the first draft is *not* a place to worry about linguistic perfection, but words do matter, and you should by no means embark on a late-stage revision of your work without ready access to a good dictionary—preferably together with a concise, practical guide to grammar. A first-rate dictionary has much more to offer than proper spellings; it can also teach you a great deal about shades of meaning, usages, and word origins.

Usage Consider, for example, the word "apparently", employed by someone who really wished to signal the status manifest in "seemingly" or "ostensibly"; or "multiple" used where "manifold" would be more suitable; perhaps a "comprised" inserted as a misguided substitute for "composed". The author who carelessly (or lazily) tolerates mistakes of this kind is entrusting to his or her words responsibilities they are incapable of fulfilling, and at the same time undercutting the legitimate reasons for those words' existence. Speaking generally, an essential precursor to enhancing the clarity of one's writing is analyzing very closely, and thinking clearly and intently about, what one actually wishes to communicate. This exercise should entail extensive recourse to a dictionary, as well as frequent consultation of a thesaurus in the determined quest for the most applicable word in a specific situation. Gaining added insight into the etymology or origin of a somewhat unfamiliar word can also be surprisingly beneficial. In addition, no matter how much confidence you have in your writing you are almost certain to profit from perusing one or more of the many books with sections dedicated specifically to frequently misused words (e.g., that old standby, STRUNK and WHITE's *The Elements of Style*, first published in 1957, and now in its fifth edition).

"Fillers" It is astonishing how much useless verbiage one finds strewn through otherwise laudable work from well-educated and experienced writers. Classic examples include "thens" and "thats" introduced needlessly in contexts like

Ex 4–1 a   If it's really necessary, *then* we should ...
It seems *that* this is the best solution ...

An even more striking example now plagues modern conversational English, especially as practiced by the young in America: monotonous, recurrent, and senseless interjection almost anywhere of the seemingly innocent word "like" (e.g., "We went there—like—three weeks ago ...").

Redundancy also deserves mention here, as in

b   potential hazard (a "hazard" is already a "potential danger")
of a critical nature ("critical"!)

To this might be added the use of wordy clichés like

c   "last but not least"
"at this point in time"

A loosely related misdemeanor, one especially objectionable in a scientific setting, is the casual sprinkling around of words specifically intended to supply *unusual* emphasis or weight, or to attract more than passing attention, but which are so worn through mindless repetition that they instead engender doubt about how seriously a piece of text should be taken; e.g., words like

d especially, very, most, above all, extremely, extraordinary, outstanding
  ("This exceptionally outstanding procedure ...")

Adjectives Decorative adjectives in general have little place in scientific text. Your analytical balance is, and should be, just that: an analytical balance—not "the trusty old analytical balance".

Moving on to another common but unfortunate flaw related to word choice and verbosity, consider the following sentence:

Ex 4–3 X represents a threat to the avian population.

The verb "represents", in a setting like this, is mere "padding". Think about it. The "X" referred to is not serving as a *representative* of anything: "X" simply *is* a threat—to (expressed more directly)—"birds".

"Bureaucrat-speak" The preceding example illustrates a wide-spread tendency often ascribed explicitly to bureaucrats: useless complication of the inherently simple. A comprehensive list of offending hackneyed expressions would be very long indeed. For starters, consider miscreants more or less analogous to "represents", such as "corresponds to", "constitutes", "can be regarded as", ... and so on and on. The case in Ex 4–3 serves to point up a second sin as well: "substantivization" or "nominalization" (crudely expressed, "nounification"), or resorting slavishly to a noun structure when what is really called for is a verb. Why not simply affirm here that "X threatens birds"? By definition, verbs have the task of heralding some sort of action, and it is therefore well-chosen verbs that are most adept at conferring a sense of motion and dynamism on a piece of text. Don't allow yourself to be unduly influenced by those who write as if they were only able to think in terms of objects and concepts, causing them to concoct wordy expressions like

Ex 4–4 ... methods for the preparation of
  ... served as a demonstration of
  ... if problems arise in the completion of
  ... perform a measurement of

rather than settling in the first case for the more natural "ways to make" or, in those that follow, simply "showed", "have trouble completing", and "measure", respectively. This addiction to erecting detours by way of awkward and verbose nominative forms is one of the most common annoyances infecting writing for the sciences (cf. C 4–3 and C 4–8). The fact that it is far less conspicuous in oral discourse (even among scientists!) underscores its fundamental artificiality. Consider one further illustration:

Ex 4–5a … an increase (decline, intensification, …) was observed (recorded, noted, …) in the course of …

all of which amounts to nothing more than a pompous substitute for the obvious and far more economical

b … increased when …

Active and passive voice

We urge you in general, as often as you can, to incorporate lively verbs into your writing. This admonition of course does not apply to inflated stand-ins for "to be" like "occurred", "happened", and "transpired", but encourages deliberate inclusion of verbs that *describe* or *characterize* actions. Try also to take as much advantage as you can of sentence structures rooted in the active voice, which is the most welcoming environment for "words of action". Scientific writing admittedly poses a troublesome dilemma in this regard: if a major objective is to minimize one's apparent personal influence over results to be described, then a powerful incentive exists to write extensively from the point of view of the things that were affected—i.e., to revert to the passive voice. Few would dispute the assertion, however, that passive constructions rarely contribute in a positive way to descriptive prose. The basic problem you must of course face is an outgrowth of the general preference in scientific text for eschewing personal responsibility, with most authors going to great length to avoid first-person—"I" or even "we" formulations. The fundamental quandary is clearly one for which there can be no strictly linguistic resolution.

Adjectives

Nouns are not the only culprits conspiring to squeeze verbs out of scientific text: adjectives often share the guilt. Consider the illustrations in Ex 4–6, each of which is accompanied in parentheses by a more active potential replacement:

Ex 4–6 … is dependent upon (depends on)
… is equivalent in its behavior to (acts like)

Word combinations

Careful writing entails paying attention also to the way words behave in combination. Should one in a particular situation say "compared to" or "compared with", for example? Bringing up the useful WorldWideWeb site

http://dictionary.reference.com

and typing in the word "compared" produces the following helpful response:

Usage Note: Compare usually takes the preposition *to* when it refers to the activity of describing the resemblances between unlike things: *He compared her to a summer day. Scientists sometimes compare the human brain to a computer.* It takes *with* when it refers to the act of examining two like things in order to discern their similarities or differences: *The police compared the forged signature with the original. The committee will have to compare the Senate's version of the bill with the version that was passed by the House.* When *compare* is used to mean "to liken (one) with another," *with* is traditionally held to be the correct preposition: *That little bauble is not to be compared with (not to) this enormous jewel.* But *to* is frequently used in this context and is not incorrect.

If you find yourself in doubt in a case like this, and you lack access to an authoritative reference source, you may have no choice other than trusting your instincts, but if so, at least try to be consistent!

Simply taking the trouble to concentrate your full attention on what you have written can be surprisingly rewarding. For example, it should help you identify and edit out unworthy, illogical pronouncements like

Ex 4–7   … the presence of X in the eluate was measured.

Think about it: the mere *presence* of something isn't a basis for "measurement"! The verb selected here ("was measured") is not a valid match for the subject ("presence"). A reader could even be forgiven for wondering if this sentence—as written—is a clumsy attempt at concealing something, possibly that even the author is a bit uncertain precisely what he or she really did during the experiment. Was *notice* simply taken that X was present in the eluate, or was the *concentration* of X in the eluate established? It might even be that a conscientious effort was made to measure the *change* in concentration of X as a function of *time*. Scientific progress is crucially dependent on many forms of precision, not the least being precise reporting.

Terminology   In any technical discussion it is important to confine oneself to using "official" terminology, and to maintain strict consistency in those relatively rare situations in which more than one expression might be considered permissible. In everyday speech or journalistic prose there is no harm at all in unleashing a varied and colorful vocabulary—in fact it is refreshing!—but variety becomes a harbinger of confusion in a scientific context, and is devastatingly incompatible with the objective of unambiguous documentation.

Sentences   The smallest true "message unit" in a document is not the word, but rather the *sentence*. Sentences (like words) can vary greatly in length, with there being no such thing as an "optimum length". Indeed, this particular realm is one in which variety is *desirable*. To the reader, a string of similar sentences, especially if similar in respect to both length and structure, is sure to feel monotonous. An assortment of sentence lengths and architectures is far better, because this sort of writing has a much greater chance of making an impact, and will certainly seem less static. At the same time, avoid sentences that are unnecessarily complex and entangled, or run on interminably. Problems of the latter sort often have their origin in "that" constructions:

Ex 4–8 It is well established that ... (of course)
One can assume that ... (presumably)
From this result it follows that ... (thus, hence, therefore)
We must not take it for granted that ... (unlikely)

Main thoughts in main clauses Note that the examples above are followed in parentheses by a word or words with the potential to head the offending sentence off in a more promising direction. Sometimes a single well-chosen word is able to replace an entire phrase, simultaneously eliminating (in most cases) the need for at least one punctuation mark. Equally important, a remodeling in this sense often permits the true message of the sentence to migrate from an awkward subordinate clause to the main clause, where it belongs. Thus,

Ex 4–9a We have shown that ...
With this experiment it could be established that ...

can be efficiently transformed into the more effective

b As we have shown, ...
As this experiment demonstrates, ...

Punctuation There is, by the way, no hard and fast rule that every sentence must end with a period. An occasional exclamation point, colon, or "long dash" (equivalent to a pause) can significantly loosen up an otherwise dry piece of discourse. Other possibilities include coupling two main clauses together with the aid of a comma and an "and", or once in a while posing a question. Take stock of your options, and exploit them fully!

Paragraphs A *collection* of sentences from a paper becomes the basis for a *paragraph*. It is advantageous to ensure that paragraphs, like sentences, vary in length. An especially important sentence can even be allowed to constitute a paragraph all by itself. In general, paragraphs should probably not exceed about a third of a page, equivalent to perhaps four to eight sentences. Paragraphs make an important contribution to the intellectual structure of a document, quite apart from the fact that they break up an otherwise "endless stream of text", which is unsightly and inflicts too much organizational burden on the reader.

Paragraph content Structuring a document as a series of coherent paragraphs forces the author to sort out and present his or her thoughts in a logical fashion. That is to say, a proper paragraph is dedicated to exploring a single subject or thought, which in principle could easily be articulated in a brief descriptive title. (In a sense, the keywords we have set in the margin in this book might be viewed as just such "titles" with respect to the adjacent paragraphs.)

Topic and bridge sentences

The first sentence in every paragraph warrants special attention. One of its principal functions is to let the reader know what the ensuing discussion is about, for which reason it is often referred to as the"topic sentence". The corresponding topic is then pursued in sentences that immediately follow, with the final sentence in the paragraph so crafted that if supplies a bridge to the next paragraph. As an example, consider the material above adjacent to the marginal notation "Paragraphs". The opening sentence contains in a prominent spot (i.e., at the very end) the keyword in question, which is what the passage is all about. (Note that the preceding sentence had quite a different theme.) This sentence in fact has the assignment of introducing what is meant by a "paragraph". There follows a brief consideration of paragraph length and structure. The final sentence subtly points the way to the next topic considered: through explicit mention of organization, and concern expressed on behalf of the reader.

The organizational factor is especially prominent in paragraphs from the "Experimental Section" of a thesis. Here each paragraph typically begins with a declaration of the goal of one particular experiment, perhaps in the form of a sentence fragment rather than a grammatically complete sentence:

Ex 4–10   Influence of X on Y under conditions Z.

Next will appear a set of procedural statements ("… were added …", "… is heated until …"), culminating in a closing sentence that indicates the extent to which the envisioned goal was in fact reached; e.g.,

Ex 4–11   … where the resulting liquid phase was interpreted to be the desired enzyme solution.
… which in turn led to a value of X for the permeation constant.

A somewhat detailed consideration of experimental descriptions has been presented here rather than in Unit 14—where it arguably would fit more logically—mainly because of the central role structural and organizational issues play. Thus, each paragraph in an Experimental Section is expected to encompass one discrete set of procedures or observations, with clearly defined starting and ending points.

Interestingly, you may find that your writing style in general will benefit from frequent practice at discussing scientific experiments in a formal way!

Paragraph titles Occasionally a keyword or two summarizing the role of a particular paragraph (in effect, a "paragraph title") is actually tacked on as a sort of preface to the text of that paragraph: separated from the text by a period or a colon and set in either boldface or italic type. Stereotypical examples of keywords in this sense, applicable specifically to experiment write-ups, would include

Ex 4–12 **Objective:** Text text text
Procedure: Text text text
Analysis: Text text text

Analogous labels tailored to a mathematical treatise might be "Proposition", "Solution", and "Proof". It was once customary, by the way, always to end formal mathematical proofs with the letters "q.e.d." from the Latin "quod erat demonstrandum" ("which was to be proven")—a fitting way to mark the successful conclusion of a challenging logical endeavor.

From this point it is but a small step to the more comprehensive subdivisions characteristic of most lengthy documents: subsections and chapters (see Units 3 and 8). These embrace whole families of paragraphs, and are always introduced by titles in the strictest sense. It is highly desirable to ensure that a title of this sort appears at least every five or so pages throughout a work. Should this *not* be the case it probably means you have not devoted sufficient attention to structure, in that your subdivisions provide insufficient evidence of organization. (Another factor to bear in mind in this regard is that *cross-referencing* between subdivisions—by section number or title—works effectively only if the reader interested in pursuing such a lead is not confronted with the prospect of a time-consuming search.) Subdivisions typically encountered in scientific theses are the subject of detailed discussion in the various units that follow.

C 4–1 Look in the literature and in your own writing for examples of unnecessary nominalization (of the "it was subjected to analysis" type), and recast those you find in ways that take advantage of active verbs.

C 4–2 Revise the following phrases such that the key information is supplied by verbs.
... subjected to investigation.
... cooled until it was transformed into the solid state.
... showed a significant improvement.

C 4–3   For each of the following statements, suggest changes that would be constructive:

The corresponding evaporation process should be brought to a conclusion as rapidly as possible.
A determination of X was then initiated using Y.
Preferential coupling of A with B is favored in this case.
The idea of expansion of U with V was subjected to serious consideration.
The general suitability of such a procedure was introduced for discussion.
Analysis of Z was accomplished spectroscopically.

C 4–4   Rewrite the sentence below in such a way as to eliminate the "that" construction.

The fact that at a pH > 8 the activity of X is dramatically altered led us to …

C 4–5   Why would a remark like

… was isolated in a most satisfactory yield.

be inappropriate in a scientific presentation?

C 4–6   What is wrong with the following sentence, and how might the problem be solved with one minor change?

The nutrient content of barley differs from corn.

C 4–7   Rewrite the following passage so it becomes easier to interpret.

The attached experimental procedure, developed over the last decade under the auspices of the Department of Defense, provides an approach to establishing the corrosion resistance of structural components based on XY after storage in a warehouse not equipped with moisture control devices. Since it was known that temperature has a major impact on the rate of surface oxidation of XY plates submerged in water, we maintained temperature constancy with air inside an experimental chamber designed to simulate the warehouse storage space with the aid of a Z control. Results at the three different temperatures selected for study are supportive of the conclusion that corrosion susceptibility will only be manifest if the air constituting the surrounding environment contains acidic components (sulfur dioxide, nitrogen oxides).

C 4–8   Compose improved counterparts for the sentences below.

… The assumption has in the meantime become quite widespread that recovery and recycling of used chlorinated hydrocarbons can be enhanced dramatically through rigorous avoidance of admixing with other waste solids or solvents.

… At the same time, with respect to both computer hardware and software, new trends and substantial increases in efficiency have been brought to the attention not only of distributors but also consumers. In place of the anticipated simplification for the user, what has instead made an appearance is a need for even greater specialized knowledge of a great many systems in order to permit judgement to be passed not only on potential utility but also sensible application.

… Of major significance for successful synthesis of this amino acid is the necessity of monitoring carefully at all times the pH level of the reaction mixture in addition to the dropwise rate of introduction of the diamine starting material, since otherwise the side-reaction (Equation 12) results in a profound decrease in the yield of desired product.

C 4–9   What is wrong with the sentence fragment below, and why?

There was then added 10% newborn calf serum and 1% essential amino acids.

C 4–10   Devise more polished versions of the following:

… the four years previously discovered method …
… the inconspicuously proliferating avian flu virus …
… the advancing temperature …

5 Writing Techniques

- This unit offers technical suggestions regarding the preparation of text for your thesis, and also on how the final document should look.

- Taking these observations to heart will help you devise an optimal strategy for developing a first draft and then transforming it into a finished product.

Q 5–1 What are some of the advantages of word processing over working with a typewriter?

Q 5–2 How should margins be set in a draft, and where do page numbers belong?

Q 5–3 How much blank space is appropriate above and below headings of various kinds?

Q 5–4 What should you keep in mind as you scrutinize your work for the last time prior to final printing?

1.4.2, 2.3, 5.1 to 5.3, 5.5

Preparation In order to prepare a good first draft of your thesis as quickly and efficiently as possible it is important that you have before you a carefully constructed outline to serve as a guide (see Unit 3). Relevant laboratory notebooks and interim reports should also be close at hand, together with descriptive notes, photocopies, or even originals of background literature you propose to cite. The latter are critical for monitoring the validity of assertions your thesis will contain regarding what is already public knowledge. Moreover, you will want to be able to include proper literature citations in your very first draft. Finally, it is a good idea already at the outset to assemble a complete set of rough sketches for graphic elements you propose to include, as well as spectra and other materials of potential interest as figures.

Rough draft | Your first attempt at text preparation will culminate in a "rough draft", which will then become the basis for the first of numerous "revised versions" (first, second, etc.).

Word processing | No one today should even consider developing the manuscript for a document the size of a thesis*) by any method other than computer-based word processing, which permits one to undertake corrections, modifications, deletions, insertions, and text rearrangements with astonishing ease. Modern word processing offers the writer many additional advantages as well:

Database links | – A full-featured word processor can be linked in powerful ways with a database containing all the bibliographic information at your disposal, permitting transfer of source identifications directly into the evolving text and thereby avoiding from the beginning all risk of typographical errors.

Footnote management | – Even if you choose not to couple your text in a sophisticated fashion with a comprehensive literature database it will still be worth your while to allow the word-processor to *manage* your bibliographic references—which ultimately will appear in the form of footnotes (or endnotes). This permits sequential citation numbers to be assigned automatically, with subsequent updating as necessary to reflect alterations to the text. A "footnote manuscript" developed in this way could, in principle, also become the basis for a separate literature section in the finished thesis.

Spelling, hyphenation | – Word processors offer a number of valuable convenience features as well, such as spellchecking and end-of-line hyphenation capability. The latter is especially important in preparing "justified" text, but it can be useful even with "ragged-right" (non-justified) typesetting, since it helps smooth the profile along a document's right-hand margin.

Style sheets | – Another powerful component of nearly all word processors—and one far too often ignored!—is format management through "style sheets". These effortlessly ensure consistent treatment throughout a document of frequently recurring elements such as body-text paragraphs, passages indented in special ways (like block quotations), free-standing material (e.g., "displayed" equations), headings at various levels, and tables, figure captions, and footnotes.

Formulas | – Word-processor text files gladly play host to professional-looking mathematical and/or chemical formulas prepared independently with the aid of appropriate auxiliary software.

---

* Note that a typical master's thesis encompasses something like 40–80 pages, a doctoral dissertation perhaps 100–300.

Graphics

– Similarly, almost any imaginable graphic element (e.g., technical diagrams)—whether scanned or assembled from scratch with a drawing program—can also be incorporated into a word-processor file for display, at any desired scale, anywhere you wish. Nevertheless, it is well to bear in mind the even greater flexibility (also in other respects!) inherent in true page-layout software.

Mechanics of revision

Once you have completed a rough draft of your thesis, as a word-processor file, the extremely demanding and time-consuming—but indispensable!—chore of editing and revision can begin. Consistent with the way one takes on other intensive reading assignments, we strongly recommend that thesis revision be carried out almost exclusively on the basis of *printed* output, not while gazing at a computer monitor. Proposed changes are rapidly but legibly noted in pencil on the printout for computer input later as a separate operation, via the keyboard. This two-step regimen might sound inefficient, but experience strongly supports our conviction that it produces much better results than direct correction of a computer-screen image.

With text double-spaced on standard letter-sized sheets of paper (one side only!), ample space will be available between the lines for most corrections. We nevertheless encourage the additional expedient with early drafts of leaving extra-wide right margins (three inches, say) to facilitate more extensive changes. A standard one-inch margin should suffice on the left. Should you find yourself making two or more editorial passes through a single piece of copy you will find it helpful to alternate the colors of changes from one reading to the next.

At some point, so many changes will have accumulated that the text you are correcting becomes impossible to read with any degree of fluency, a signal that the time has come to transfer the amendments to your digital file in preparation for printing a fresh copy of the manuscript—for the next round of editorial polishing!

Multiple versions

The process described will of course lead to a whole series of versions, each of which should be clearly labeled through dated page headers like

Ex 5–1   Version 3, 3/1/05

The headers should also indicate page numbers, preferably near the right margin where they will be easy to see. (Word-processing programs typically support custom headers and footers that automatically fill in the correct page number and current date.) At the end of each working session, be sure you prepare backup copies of your latest efforts. For safety's sake, at least one up-to-date copy should be stored in a separate secure location.

Revision Whereas your primary concern during development of a rough draft must be establishing a logical and comfortable flow for the key ideas, in the editing and revision stages you will focus mainly on linguistic and stylistic issues. Does every sentence express precisely what you intended it to say? Will all your thoughts prove intelligible to the prospective reader, who will inevitably be less familiar with the material than you? Have you successfully removed all threat of misinterpretation?

Checking the formalities As the time approaches for preparing a "final" copy, begin refining your page layout so that it comes increasingly to resemble the model dictated by your particular university's thesis instructions. Also check carefully to be sure that miscellaneous formal matters have been properly attended to:

- Are all figures and tables suitably anchored in the text (cf. Units 20 and 21)?
- Has each figure and table been assigned the correct number?
- Is there strict uniformity with respect to the presentation of headings (e.g., consistent type style and size for every heading at each of the various hierarchical levels)?
- Is the right amount of blank space present before and after every heading (see below)?
- Are type assignments correct for the various kinds of text elements (e.g., body text, tables, footnotes)?
- Is line spacing satisfactory throughout, not only in body text but also in tables, footnotes, figure captions, etc.?
- If foreign elements still need to be added (such as pasted-in figures), is appropriate space available in each case?

Last corrections We urge you to share what you hope is the "final version" of your manuscript—one you are tempted to print for formal submission—with one or two trusted colleagues for their critical review and comment (and of course with your advisor as well, assuming this is accepted practice in your research group). You will as a result almost certainly be apprised of the need for at least some further revision, which should however pose no major problems, since—thanks to word-processing—your "electronic document" is always subject to easy modification.

Final copy; printing and binding Now serious attention must be devoted to a true "final copy", obviously incorporating polished versions of special elements like figures, formulas, and tables. This final version should be prepared with a laser printer, or at least a high-quality ink-jet printer with resolution of 300 dpi or greater.

Printer output would in most cases then be photocopied to yield the actual document destined for binding and submission, an advantageous detour in that it facilitates preparation at the same time of the additional clean copies you are likely to want later.

Type fonts   One of the many advantages of modern word-processing systems is their support of a vast array of type fonts, sizes, and styles, including strictly proportional characters (either with or without serifs; cf. Fig. 5–1a, b) as well as non proportional letters and numbers that resemble those issuing from a conventional typewriter (Fig. 5–1c). The distinction here lies in the fact that with a *proportional* font each letter lays claim to an amount of horizontal space consistent with that symbol's inherent width. In a non-proportional (or "monospaced") font, by contrast, all characters are of equal width, and they are spaced uniformly. In final copy we recommend you make at most sparing use of both non proportional fonts (Fig. 5–1c) and fonts lacking serifs ("sans serif" fonts; Fig. 5–1a). Indeed, proportional fonts elaborated with serifs (for easier reading) are today taken almost for granted in running text for formal documents like theses.

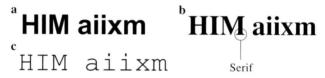

Fig. 5–1. **a** Sans-serif proportional font (e.g., Arial); **b** Proportional font embellished with serifs (e.g., Times); **c** Non proportional font (Courier).

Treatment of the right margin   It is also important to consider one additional nicety that contributes significantly to the professional appearance of a polished, "finished" work: "justified" text, in which one arranges for all paragraphs to have razor-sharp, carefully aligned right edges. This is achieved by instructing the word processor meticulously to adjust as necessary the spaces provided between words. The alternative is a "ragged right" margin, interpreted by many readers as a sign of a rather informal document, or a "work still in progress".

General text characteristics   Assuming you select a dignified, proportional type font (e.g., "Times"), and specify fully justified text, the final version of your thesis will look like it came from a professional printing establishment—assuming of course that you also adhere strictly to certain special rules that have evolved over decades with respect explicitly to the typesetting and typography of scientific material. (These are summarized in a general way in Units 18 and 19).

With standard letter-size (8.5″ × 11″) paper oriented vertically ("portrait format"), a normal text block from a typical thesis will be about 6″ wide, with 1″ margins everywhere except on the left, where 1.5″ is preferred (to facilitate binding). Note that this means the "center" of a particular line of text will actually lie 1/4″ to the right of the centerline of the page itself (cf. Fig. 5–2).

Fig. 5–2. Typical text margins and page-number placement with a standard 8.5″ × 11″ page for a thesis.

Pagination   Pages are generally numbered at the top, centered above the text area, with the numbers in turn flanked by dashes (preferably "en-dashes"), as, for example:

Ex 5–2   –37–

This information often appears within 3/4″ of the top of the page, with blank space below it equivalent to at least one line of text.

Type size, line spacing   It is important for appearance sake that lines of running text be spaced evenly, but also widely enough so that there can be no chance of overlap between characters in adjacent lines. In dealing with this requirement be sure to take fully into account the vertical displacements often compulsory in mathematical expressions. Assuming, for example, that the indices so ubiquitous in scientific work (i.e., subscripts, superscripts) are set in type ca. 30% smaller than body text, "line-and-a half" spacing will normally serve the purpose, although word-processors make it extremely easy to assign finely-tuned custom spacings where required, measured directly in "points". In keeping with the suggestion above, indices associated with 12-point body text might be set in 8-point type, with a line spacing of 18 points. Just as is usually true with professionally published works, type utilized for figure legends, table content, and footnotes should be noticeably smaller than that assigned to body text: 10-point, for example, with 12-point body text.

An interesting rule always observed by professional printers—but often ignored by amateurs—specifies that no page should ever begin or end with a single-line fragment of a paragraph (referred to respectively as an "orphan" or a "widow"). Sophisticated word processors (e.g., WORD) include provisions for automatic compliance with this rule in most cases, but you may still find it necessary on occasion to intervene manually by adding or deleting one or more words in some line, or introducing a forced page break.

Heading placement We recommend that headings of all sorts be set flush against the left margin. This is not only easier to achieve than the alternative of centering, but it also avoids the companion obligation to center all of one's tables and illustrations (together with their captions), elements traditionally treated in the same way as a document's headings. Tradition also provides standards for the vertical placement of headings: space reserved *above* a heading should always be 1.5 to 2 times as great as that separating the heading from subsequent text (which might of course be another heading).

Ex 5–3 **1 Heading**

1.1 Heading   ———— 1.5:1 to 2:1

1.1.1 Heading

Text Text Text Text Text Text Tex Text Text Text Text Text Text Text Text Text Text Text Tex Text Text Text Text Text Text Text Text Text Text Text Text Text.

Visual characteristics of headings With respect to typographic distinctions applied to the headings at various levels, published works often rely exclusively on different type *sizes*; e.g., 24-point, 16-point, and 12-point type for headings at the first, second and third hierarchical levels, respectively. Another possibility is maintaining a constant *size* for all heading type but varying the type *style*; e.g., **bold** for first-level headings, ordinary (roman) for second-level headings, and *italic* for third-level headings. Some elect to use for headings a different font altogether from that of running text: perhaps an unadorned sans-serif font of an appropriate size and style.

Single-side printing Pages in theses and dissertations normally contain text on one side only, which of course means that in a bound format all "left-hand pages" will be blank.

University guidelines It is absolutely essential that you ascertain from the appropriate authorities at your own university—in a timely fashion!—precisely what their expectations are regarding any thesis you submit: number of copies, binding style, paper characteristics, page layout … the list can seem endless, but *none* of the published "guidelines" (i.e., regulations!) can safely be ignored. It would in fact be a good idea prior to final printing to show a few sample pages of your work to the party responsible for monitoring such things at your institution, and to examine closely on your own a representative selection of recently submitted works analogous to yours.

ETDs But wait! Are you quite certain that your institution in fact insists upon *being* presented with a traditional printed and bound thesis? Or is that merely one option? Many long-held practices are today in a state of flux in academia, and at many schools "electronic" theses have recently been declared acceptable—or even preferred. This development is to some extent a consequence of a meeting organized in 1987 by University Microfilms, Inc., and attended by interested parties from a wide range of universities. University Microfilms has for decades assumed primary responsibility for "publishing" academic theses in general, and they have long sought ways to facilitate the broader dissemination and recognition of theses as valuable primary information sources. The most important outcome of the 1987 meeting was formal acknowledgement of what have since become known as "ETDs": *E*lectronic *T*heses and *D*issertations. Many institutions now sanction the preparation of theses consisting exclusively of digital data files, provided the files meet certain strict standards. It is of course impossible to say how soon or even *whether* such an option will be available everywhere, but a distinct trend is apparent in that direction. What has already become clear is that:

– From the technical, archival, and distribution standpoints, digital (electronic) theses offer many distinct advantages, and
– Certain types of material can obviously be made part of a "document" of this type much more easily and satisfactorily than with the traditional book format.

Given the rapid evolution of attitudes and practices in this area it would be premature for us to go into great detail here about the preparation of digital theses, although we encourage you through one of the exercises at the end of this unit (C 5–6) to acquaint yourself with some of the issues involved. An especially interesting and relevant development, spawned in the United States, is a consortium known as the Networked Digital Library of Theses and Dissertations (NTLTD), with nearly 150 affiliated universities worldwide.

Despite widespread agreement regarding certain basic principles, individual universities continue to promulgate their own detailed regulations for how "electronic theses" are to be prepared and submitted. Most provide some level of support through various helpful resources, including software. In general, one would be expected to assemble and submit electronic thesis content in the file format "PDF" (Portable Document Format), an alternative developed and introduced by Adobe Systems in the early 1990s which has in recent years come to be regarded as something of a standard worldwide. PDF files are prepared as combinations of other widely used document-file types (e.g., WORD, EXCEL, jpg), typically with the help of a compilation program like Adobe ACROBAT. PDF files make it possible to display on any computer platform a document that looks precisely as it was envisioned by the author, with the same pagination, type fonts, etc. Often (but not necessarily), access to PDF files is secured through the free software package Adobe ACROBAT READER.

Other technicalities associated with the writing of scientific theses are reserved for treatment in depth later, in particular footnotes (Unit 17), quantities (unit 18), equations (Unit 19), and tables (Unit 20).

C 5–1   What are some of the most important things to remember as you set out to prepare a rough draft of your thesis?

C 5–2   Describe differences to be anticipated between a rough draft, an edited version, the final draft, and the final copy of a dissertation.

C 5–3   List several advantages of word processors relative to typewriters.

C 5–4   How can one differentiate text headings at various hierarchical levels (e.g., first order, second order, etc.)?

C 5–5   How much free space should be provided before and after a heading?

C 5–6   Cite some of the advantages associated with the preparation, submission, archival, and public distribution of electronic theses. (Surely you can come up with at least *eight*!)

Part II
The Components of a Thesis

A word of introduction

The thesis structure shown below, and presupposed throughout Part II ("standard structure"), in many cases proves to be a very convenient one, and is certainly fairly common, but it is by no means absolute.

- Title page
- Preface, Acknowledgments
- Contents
- Abstract
- List of Symbols
- Introduction
- Results
- Discussion
- Conclusions
- Experimental
- Bibliography
- Appendices
- Remarks
- Vita

Certain chapters in the list are sometimes situated elsewhere in the thesis, assigned different titles, combined with others, or even omitted entirely. You should also not rule out a completely original structure, at least not before closely examining your particular institution's requirements.

6 Title, Title Page

- We deal in this unit with basic considerations applicable to selecting a title for your thesis, as well as the layout of the title page.

- You will discover that a title page is in some ways reminiscent of a business card, and spending some time thinking about this page's structure should give you a better appreciation for its function.

Q 6–1 What is the purpose of a title, and what should it address?

Q 6–2 To what extent should there be a relationship between "keywords" relevant to one's thesis work and the title of the corresponding document?

Q 6–3 What factors should be taken into account regarding the length of a thesis title, and when should you consider dividing a proposed title into a main title and a subtitle?

Q 6–4 Is it ever permissible to include symbols and abbreviations in a title?

Q 6–5 Apart from the official title of the document, what other information belongs on a title page?

Q 6–6 Should the title be centered on the title page, or left-justified? What about other information present?

Q 6–7 Are there standards applicable to spacing of the various title-page elements?

Q 6–8 How can the thesis title be caused to stand out from other information on the title page?

 2.2.2, 5.5.1

Characteristics of a good title

The first principle to keep in mind is that the title for a thesis should be kept as short as possible consistent with clearly and comprehensively conveying the theme of the work. Rather like an extremely compact summary, a title should provide insight into the subdiscipline to which the study relates, and at the same time reveal something about the chief goals, methods, focus, and/or results. Even a title that has been divided into a main title and a subtitle should not in its entirety exceed roughly 200 characters (or 4 lines), including spaces, which is the equivalent of about 25 words.

In developing your title, try to incorporate as many keywords specific to your project as you can. These might be derived from the topic under investigation, techniques you employed, noteworthy accomplishments, or any combination thereof—in other words, the "What?", the "How?", and perhaps even the "Why?" of the work. A title like

Ex 6–1 a A New Method for Analyzing Fluoride-Containing Solutions

conveys virtually no meaningful information, and would be much better replaced by something more explicit, like

b Automatic Photometric Fluoride Titration, with Thorium Nitrate and Alizarin-S as Indicator

The ideal title consists of no more than about ten or twelve words. If you find this standard too restrictive, try solving the problem by appending a subtitle, but make sure the most important information still falls within the main title. A subtitle should be limited to furnishing supplementary thoughts that clarify the title, or delimit it in some way. Often a subtitle functions mainly as a device to shift some critical term forward within the main title, because for documentation purposes first words play a uniquely important role.

If one thinks about a thesis (or, for that matter, a journal article) in its entirety as a *message*, then the title represents the corresponding *address*; like the key components of the address appearing on an envelope, a document title must be both precise and comprehensive in order to ensure that the message in question actually reaches the party or parties for whom it was intended.

An overly verbose title, like

Ex 6-2 a The Chemical and Physical Properties of Perfluoroalkane Sulfenyl Fluorides as Products Derived from Treatment of Perfluoroalkane Sulfenyl Chlorides with Silver Fluoride, and their Characterization through IR-, Raman-, and ^{19}F-NMR-Spectroscopy

stands a much better chance of success if reformulated into

b Perfluoroalkane Sulfenyl Fluorides from Perfluoroalkane Sulfenyl Chlorides and Silver Fluoride: Chemistry, Physical Properties, and Characterization through IR-, Raman-, and ^{19}F-NMR-Spectroscopy

Special care must be taken in the choice of a subtitle (in the example above, everything after the colon) to be sure that its relationship to the main title is absolutely unambiguous. Subtitles are sometimes expressed in the form of entire sentences, complete with verbs, especially if framed as questions:

Ex 6-3   Is the Effect of the XXX-Factor Dependent Upon YYY?

When a subtitle begins on the same line as the end of the main title (cf. Ex 6–2 b; this stylistic convention is common in professional journals), the main title invariably concludes with a colon.

Titles should never contain abbreviations in the everyday sense, although common technical abbreviations or acronyms like

Ex 6–4a   IR   for infrared
NMR   for nuclear magnetic resonance
DNA   for deoxyribonucleic acid

are usually considered permissible. Less familiar technical abbreviations of special importance to your particular work should be elaborated for clarity; e.g.,

b   … by Flame-Ionization Detection (FID)

Special symbols   Insofar as possible, titles should be kept free of special symbols and formulas, although strict adherence to this rule can become impractical in cases like

Ex 6–5   … with ω-Substituted Octenoic Acids …
… of the Type $Al_2Si_3Lu_3X_3$ …

Bland expressions of the type

Ex 6–6a   Experiments on …
Investigation into …
Attempts to …
Results of the …
Determination of …

make no useful contribution whatsoever to a thesis title, and should be banned; even if such an element does for some reason seem to be called for it should be relegated to a subtitle. A main title is also not the place for qualifying words like "comparatively", "theoretically", and "historically", which should also be restricted to subtitles. Try instead to come straight to the point, taking advantage of opening phrases comparable to

b   Ligand-Exchange Reaction of …
Enzymatic Synthesis of …
Optimization of …
Optical-Acoustic Determination of …
Rheological Properties of …
Circular Dichroism in …
Microfiltration through …
Ion Implantation by …
Synthesis of … from …
Dependence of … upon …
Use of … as …

Working title   Before thesis research even begins it is a good idea to have reached an understanding with your advisor about an appropriate working title. The precise title that will ultimately characterize the resulting thesis, however, can only be formulated after all results are in hand and the document itself has been drafted.

The title of an extensive document always is assigned a page of its own, a *title page* (cf. Fig. 6–1). With a *bound* volume this is typically the first true page one encounters upon opening the cover, although it might be preceded by a single blank sheet.

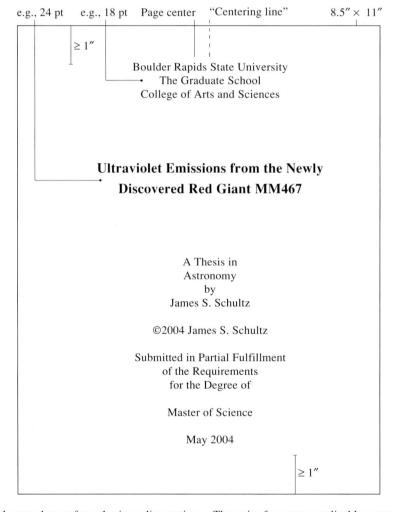

Fig. 6–1. Typical title-page layout for a thesis or dissertation.—The unit of measure applicable to type size is the *typographic point*. (In the United States, 1 pt = 0.3515 mm, although in much of the rest of the world 1 pt = 0.375 mm.)

The main title of a thesis is often quite short, especially when supplemented by a subtitle; consider, for example:

Ex 6–7a Colored Complexes:
The Charge-Transfer Phenomenon as Manifested by …

b Lead:
X-Ray Structural Study of a New High-Pressure Polymorph

Usually on a title page the main title is set in larger type than that used for a subtitle—or for that matter any other information on the page (cf. also Fig. 6–1). For example:

Ex 6–8 Hydrocarbon Permeation through Multilayered Polymer Films
Determination with the Aid of Head-Space Gas Chromatography

Another important piece of title-page information with almost every thesis or dissertation is a straightforward declaration of the context of the work, such as:

Ex 6–9a Master's Thesis
Doctoral Dissertation

The latter might be elaborated into a formal explanatory statement like

b In partial fulfillment of the requirements for the degree Doctor of Philosophy in the subject of chemistry

and perhaps accompanied by identification of the academic department and university involved ("Where?"); e.g.,

c Department of Physics
Harvard University
Cambridge, Massachusetts

Authorship In addition, a title page must also identify in the standard way the author of the work ("Who?"): a family name preceded by at least one forename, to be spelled out in full; e.g.,

Ex 6–10 George Wilson Smith

Other information may be mandatory as well, depending upon university requirements, such as a date of submission ("When?") or copyright notice.

Centering the text Format rules devised by universities and/or departments ordinarily leave little latitude for individual creativity with respect especially to the title page, and all such rules must be followed to the letter! This applies not only to information content, but also how that information is handled and displayed. Generally speaking, title-page text—without exception—is centered, where "centering" must again be understood as relative to the *type page* (or text area), not the overall boundaries of the paper. The "centering line" would therefore fall somewhat to the right of the horizontal mid-point of the page.

Pagination   Standard pagination of a thesis does not begin until the first page of the Introduction. Earlier pages in any published volume (e.g., title page, table of contents, etc.) are considered part of the "front matter", which is usually numbered separately on the basis of roman numerals. No explicit page number should appear on the title page itself, even though it would be included within the roman pagination (as "page i").

Unless otherwise specified, the beginning of the actual title of the thesis should be at least 1.5″ below top of the title page, with the last text line on the page an equivalent distance from the bottom.

C 6–1   What information is expected to appear on the title page of a thesis?

C 6–2   When is it advisable to divide a title into a main title and a subtitle?

C 6–3   Critique the following thesis titles, and suggest more suitable alternatives:

    a   Investigation of the Anaerobic Purification of Wastewater from Yeast Production

    b   Development of a New Method for In-Situ Determination of Phenols in Soil Using UV Spectrometry

    c   A Method for Establishing Particle Size

    d   Study Directed toward the Possible Utilization of a Conditioner in the Composting of Communal Waste

    e   Development of an Apparatus for Measuring the X-Vector of Y-Emissions from Anisotropic Samples, and Their Representation as a Function of Molecular Parameters According to the ZZ Formalism

    f   A Report on the Van Argen Syndrome Based on Three Case Studies

    g   In-Vitro Influences of Fatty Acids on Permeation Rates of α- and β-Glycosides

C 6–4   How could the title pages depicted in Fig. 6–2 be improved?

Calculation of Vibrational Spectra for Crystalline Benzene Derivatives

**Hexamethylbenzene
and Perfluorohexamethylbenzene**

Masters Thesis

submitted by

Herman L. Johnson

2003

Measuring the Effect of Vinpocetin
on the In-Vivo Flexibility of Erthrocytes with a New Centrifugation
Technique

Research
by
David L. Jones from Boston

2004

Fig. 6–2. Title pages that could benefit from some modification. (It is certainly not our intent, by the way, to embarrass anyone; pages shown here are strictly hypothetical, and were conceived by us for this express purpose.)

7 Dedication, Preface, Acknowledgements

- This unit introduces additional components that may precede the actual text of a thesis.

■ Here you will learn about proper form and placement for Dedication, Preface, and Acknowledgements pages within a thesis or dissertation.

Q 7–1 Where should a dedication page appear, and how should it be presented?

Q 7–2 How does a preface differ from "acknowledgements"?

Q 7–3 What purpose or purposes might a preface in a dissertation serve?

Q 7–4 How should you formulate expressions of gratitude directed toward your advisor, a technician, or your colleagues?

Q 7–5 Is a preface a proper place for a note of appreciation regarding sources of financial aid you received in support of your studies?

Q 7–6 What comments regarding the scope or specific aspects of your research belong in a preface rather than in the body of the document?

> 2.2.1, 2.2.3, 2.2.11

Dedication After the title page, but before the description of your actual work, you may wish to incorporate a *dedication page*—or even an *epigraph* (a selected quotation you regard as especially pertinent, generally from a literary source). Another page might be reserved here for a *preface*. Page numbers are not printed on dedication pages or prefaces, although such pages are included in the pagination, usually the separate roman pagination scheme reserved for front matter. The reverse side of a dedication or epigraph page is left blank even in the type of document in which pages ordinarily carry text on both sides.

Dedications are very personal in nature, and they can take many different forms. A statement in this category might, for example, be very brief, as in

Ex 7–1   To my parents and my wife Dorothy

Alternatively, one might choose to elaborate a bit:

Ex 7–2   Dedicated with gratitude to my distinguished mentor, whom I am honored also to regard as a friend: Prof. David Edward Jones.

A preface is a good place to make note of things that have nothing directly to do with the results of your research but nonetheless reflect upon its context, such as the time period in question, or the name of some sponsoring entity; e.g.,

Ex 7–3   This study, begun in January 2001 and completed in May 2005, was carried out in the Hoffman Laboratory of the Department of Chemistry, Central States University.

(Be sure to follow current accepted practices or guidelines at your own institution!)

What should *not* appear in a preface is any indication of the significance either of your topic or of results you may have obtained. Formal judgements of this sort remain the exclusive prerogative of your examining committee.

On the other hand, a preface is a perfectly appropriate place for a comment like

Ex 7–4   I wish here to express my sincere appreciation to Prof. James Devoe for entrusting me with most of the decisions related to this research project, and for lending valuable support in the course of countless informal discussions.

Ex 7–5   I am grateful to Darlene Davis of the Department of Chemistry at Western Regional University for providing me with single-crystal x-ray diffraction data.

Ex 7–6   Without the electronics expertise of George Vandekamp it would have been impossible for me to assemble the positron detector upon which the reported results depend.

Acknowledgements   If a proposed preface consists exclusively of expressions of appreciation it would probably be better to change the title to "Acknowledgements".

Either in a preface or on an acknowledgements page it is perfectly appropriate to express gratitude for financial or other support you have received. Some benefactors actually provide specific guidelines for how any such notice in a thesis or dissertation should be formulated. Representative examples include

Ex 7–7a   I wish to thank the XY Foundation for generous financial support of my studies during both 2003 and 2004.

b   I am grateful to the Ingelsbach Corporation for providing me in the course of my investigation with cost-free access to a high-resolution mass spectrometer.

Sometimes an acknowledgement takes the form of a simple statement of fact, presented as the closing sentence in a preface:

Ex 7–8 Support for these studies was provided by a development grant from the United States Department of Agriculture.

In an extended series of acknowledgements, try to avoid repetitious language like

Ex 7–9 "I wish to thank…", "I also wish to thank…", "Finally, I wish to thank…"

Should you feel the need to acknowledge a whole set of people and organizations it might be better simply to construct a list

Ex 7–11 Dr. Albert Smith assisted me with …
George Davis designed …
Mrs. Judith Elsner performed all the …

and then follow it with a "group thank-you" such as

Ex 7–12 … all of whom I gratefully acknowledge here with my sincere thanks.

Paragraphs Avoid the temptation to treat every sentence in a preface as a separate paragraph; things that are clearly related should, after all, be kept together. (On the other hand, an expression of gratitude to an advisor normally *would* warrant a paragraph of its own.)

With two-sided print, a preface always commences on a right-hand page, a rule that is in fact actually applied in general to all distinctive elements (table of contents, start of a new chapter, etc.). If a preface consists only of a single page, its reverse side would be left blank.

A preface is often concluded with a place, a date, and a name, as in the example in C 7–3.

C 7–1 Where in a thesis does a dedication belong? What about acknowledgements and/or a preface?

C 7–2 How should dedication and preface pages be numbered?

C 7–3 Suggest ways of improving the following preface, taking into account both content and linguistic issues.

Preface

The work described here was carried out in the Stearns research group of the Department of Biology at XYZ University between July 2001 and November 2003. I wish to thank Prof. J. B. Stearns for suggesting that I undertake an independent investigation in this area, and for all the support he provided. I also want to thank George Walters and the other members of the group for their contributions to many helpful discussions. Thanks also to David

Weizman, who helped me record and interpret the esr spectra. Finally, I wish to thank the Akron Institute for the generous fellowship they awarded me for the time period July 2001–November 2003.

Westville, in November 2003   Arnold Abrams

8 Table of Contents

● This unit illustrates how a set of chapter and subsection headings is transformed into a table of contents.

■ Study of the material should help you build for your thesis a good table of contents, and also see why it is important to do so.

Q 8–1 Where should a table of contents generally be placed in a thesis: at the beginning or at the end?

Q 8–2 What is the preferred system for numbering chapter titles and section headings?

Q 8–3 How should entries in a table of contents be arranged so the reader will be able to tell immediately which are most important?

2.2.5

Contents — A table of contents (often simply called "the contents page") consists of an ordered list of headings for all the chapters, sections, and subsections constituting a book (or other major document, such as a thesis), together with the page numbers that mark the start of each. Its principal functions are to indicate the internal structure of the work and to guide the reader to material of the greatest immediate interest. The table of contents is often the first page a prospective reader looks at, and it may well determine whether that book will be examined further or merely set aside.

Headings and page numbers — One can of course only prepare a complete table of contents for a complex document after the final version of the work is in hand, because all headings as listed must be strictly identical to those in the book itself, and all cited page numbers must be correct. Nevertheless, authors often find it convenient to include a tentative table of contents even in early versions of their manuscripts.

The following represents a group of typical entries from a table of contents for a thesis in chemistry:

Lengths of sections   This example apparently shows only *portions* of an actual table of contents, however, judging from the reported page numbers. Consider, for instance, the large gap between pages 39 and 78. A common rule of thumb requires that a new heading should appear at least every five or so pages, which means either Section 2.2 should be further subdivided, or else an existing set of sections 2.3, … has somehow been overlooked.

Occasionally in a table of contents the titles of chapters are highlighted relative to those of sections or subsections by their being set in distinctive type (e.g., boldface, or a larger size), perhaps in conjunction with judicious spacing; e.g.,

Aligning as above the left edges of heading text that accompanies section numbers increases the structural clarity of the table as a whole. The reader is thus able easily to recognize—from the amount of space consumed by numbers—each individual heading's relative status in the overall heading hierarchy. Proper placement of the corresponding (invisible) vertical "text starting line" obviously is a function of the physical length of the longest section number, which should be separated from associated heading text by at least two or three blank spaces.

Indentation   Document structure might become even more apparent if one were to introduce multiple levels of indentation, as illustrated in Ex 8–3, but this produces a great deal of blank space, and it gives the page a disconcertingly "busy" look.

Page numbers Instead of bridging the space between headings and page numbers with "filler" dots as in the preceding examples (which incidentally can be awkward to accomplish neatly!), numbers are sometimes printed essentially adjacent to the appropriate headings, separated from them only by a fixed number of blank spaces; e.g.,

Notice, by the way, that periods are never present after chapter or section numbers, or after heading text.

All the text included within a document is expected to be compatible with the established section-numbering scheme. Consider in this light therefore the following suspicious table-of-contents excerpt:

The page numbers cited strongly suggest that there must be text of some sort *between* the heading announcing Chapter 5 (on page 42) and that for the subordinate Section 5.1 (on page 43), since a well-designed page would never end with a heading. But that cannot be, because any such text must necessarily fall in a "classification no-man's land", which is unacceptable.

Note also that if a document contains a section numbered 5.1, this must be followed by at least one other section of the same rank (i.e., 5.2), and possibly several such sections (e.g., 5.2, 5.3, 5.4, ...). Farther down in the hierarchy, after a Section 6.3.1 there must similarly appear a Section 6.3.2.

Otherwise, introduction of a heading labeled "6.3.1" becomes pointless: a "first" should always be the forerunner of a "second"! Errors in this category are easily detected as part of a rigorous inspection of one's proposed table of contents. In the following excerpt, for example,

Ex 8–6   ...

4	Experimental
4.1	Synthesis of the Starting Material
5	Spectral Analysis
5.1	IR Spectra
5.2	NMR Spectra
5.2.1	^{13}C-NMR Spectra
5.3	Raman Spectra

...

either additional sections numbered 4.2 and 5.2.2 have somehow been overlooked, or else the designations 4.1 and 5.2.2 are superfluous (which is to say, in the absence of such additional sections it was a mistake to initiate processes of subdivision in both Chapter 4 and Section 5.2).

One should limit the numerical structuring of a set of headings to at most five discrete "levels"; in fact it is usually considered preferable to stop at three. If more extensive subdivision seems necessary, one possible solution is to break the overall work into "Parts", labeled on the basis of roman numerals. The various chapters and sections within these "parts" would then retain their original arabic designations. Chapter numbering under such a scheme should be continuous throughout the document, as illustrated in the second solution (S 8–2b) proposed for Challenge (C) 8–2, shown in the Solutions Section near the end of the book (cf. also Ex 10–5). Another option would be to create an even lower set of *unnumbered* headings, as in Ex 11–7. (One further useful "trick" is illustrated in Ex 14–1.)

Ex 8–7 depicts a (slightly edited) table of contents from a dissertation actually submitted by a student some years ago, which will serve as a comprehensive illustration of systematic subdivision numbering. Headings at the highest level here correspond essentially to the components introduced ahead of Unit 6 as representative of the "standard thesis structure". Several have then been further subdivided as appropriate. Note that the structure of the "Experimental Section" (in this case labeled "Materials and Methods" and moved to a position immediately following the "Introduction") is to some extent reflected in the related section labeled "Results".

This outline might also be regarded as one (rather good!) solution to the challenge posed in C 3–3 (Unit 3).

In the units that follow we consider in detail the principal sections included in a typical thesis—Introduction, Materials and Methods (or Experimental), Results, Discussion, and Conclusions. As noted previously, situations sometimes arise—especially in the more theoretical disciplines, such as physics—in which it is actually better to stray from these traditional guidelines. For our purposes, however, we felt it best to adhere to the accepted canon, at least in a general way (cf. the "Word of Introduction" at the beginning of Part II).

C 8–1 Suggest possible improvements in both structure and form with regard to the following excerpt from a table of contents:

C 8–2 The outline implicit in Ex 8–1/Ex 8–2 differs somewhat from the "standard" outline introduced in Unit 3. What changes have been made? How might one adapt it so that its structure would more closely resemble that of a "standard" thesis?

C 8–3 Would you judge the following structure to be acceptable for a thesis? Suggest a discipline or disciplines in which it could be appropriate. (Lower-level headings have of course been omitted.)

1 Introduction
2 Theory (or Model, or something similar)
3 Experimental
4 Results
5 Conclusions

C 8–4 Under which of the major headings in C 8–3 might the following subsections fall?

X.1 The Donnan Equilibrium
X.2 Osmotic and Sedimentation Equilibria
X.3 Disruption of the Interphase Equilibrium and the
     Dissipation of Membrane Transport
X.4 Calculation of the Dissipation Function for Membrane
     Transport

C 8–5 What purposes are served by section numbers?

9 Abstract

- In this unit we treat the goals and characteristics associated with the abstract for a thesis, and offer suggestions regarding its preparation.

- Having read this unit you should be better able to find a way of expressing the essence of your work in a *single* page, and also recognize the differences between a good and a bad abstract.

Q 9–1 Why is an abstract necessary in a thesis?

Q 9–2 Do the terms "abstract", "conclusions", "brief review", "synopsis", and "summary" all mean the same thing from the standpoint of content and form?

Q 9–3 What is typically the maximum permitted length for the abstract of a thesis or dissertation?

Q 9–4 Why are tables, figures, or formulas not allowed in an abstract?

Q 9–5 Should an abstract include any experimental details?

Q 9–6 What is the proper placement of a thesis abstract?

Q 9–7 Why do "official" standards exist for abstracts of many scholarly documents?

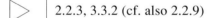 2.2.3, 3.3.2 (cf. also 2.2.9)

Abstract An abstract as part of a thesis serves the purpose of announcing—very briefly, in summary form—the goal(s) of the underlying research, methods that were employed in the investigation, and the most important results achieved.

In some ways one might argue that the "Conclusions" section of a thesis (see Unit 13) already presents a summation of the work, but as we shall see, this section is in fact quite unlike an abstract.

A good thesis is so designed that it "tells a story". It should not be perceived as a dry compilation of details, but rather as a creative account of some intellectual problem as you initially encountered it, what happened as you examined the corresponding terrain more closely, and how the scene looked at the end. The practical demands of this assignment actually leave you with much more "poetic license" than you might think. Thus, you are not expected to tell your story as it actually happened, for example, but should instead structure it in such a way that it reads well! Statements you make must of course be accurate, but the production as a whole is yours to direct as you see fit.

The abstract is intended to present the curious reader with a condensed version of your overall story. It should obviously point out all the major features of the underlying investigation, but delve deeply into very few of the details.

Perhaps the most important point for us to emphasize at the outset is that, even though the abstract usually claims a space near the front of a thesis, no attempt whatsoever should be made to compose it until the main body of the document is complete. You cannot possibly "summarize" in a legitimate way something that does not yet exist (!), and theses often evolve in totally unforeseen ways.

Preview and recapitulation

Nothing should appear in an abstract that is not also covered (at length!) in the body of the thesis itself. In other words, this section is literally to serve as a preview of that which follows, and to provide a convenient overview for the reader already familiar with your work. In a sense, the thesis abstract might be characterized as a miniaturized version of the independent accomplishments you have elected to present for evaluation.

Questions and answers

A good abstract will not leave an impression of being vague or general, but it must also not be overly detailed. Every assertion included should be very specific, and each will of necessity also be the product of a ruthless selection process. You may decide to devote one or two sentences at the beginning to background: just enough so that the reader will find the questions around which the work centers to be both intelligible and worth asking. The primary objective, however, is to sketch out what you personally have done to clarify the problem of interest, what new things you discovered along the way, and what answers you are now in a position to supply to the original questions. You may also choose to allude to implications of those answers, including cautious speculation, but again only briefly.

Condensing a document of perhaps a hundred or more pages into one single page is a daunting assignment: every word will need to be weighed carefully. "Official" abstracts of contributions to the public literature (journal articles, books) eventually become incorporated into scholarly abstract collections (e.g., *Chemical Abstracts*, *Physics Abstracts*), and for this reason their preparation is generally entrusted exclusively to specially trained editors who operate within strict guidelines. In this case, however, you must assume the responsibility yourself.

In order to ensure that an abstract is as self-sufficient as possible, and thus of value in the absence of the document upon which it is based (so that it can serve if necessary for documentation purposes), a number of criteria must be respected. In particular, an abstract—including a thesis abstract—is expected to be:

– *comprehensive:* this implies, incidentally, that you may need to make note of crucial side issues of scholarly significance;
– *precise:* facts and opinions expressed within the actual thesis must under no circumstances be exaggerated or alluded to in misleading ways (indeed, even subtle shifts in emphasis would be considered inappropriate);
– *objective:* an abstract, like the thesis itself, may include evaluative comments, but it must not seem judgemental, even if the thesis topic raises controversial issues;
– *concise:* only the most important of your results can be accommodated in an abstract;
– *intelligible:* to meet this criterion you will be obliged to restrict yourself insofar as possible to familiar terminology, symbols, and abbreviations.

Purpose Above all, the abstract of a thesis must be informative, furnishing the interested reader with a true sense of the scope and importance of the reported work, especially as this relates to the reader's own interests. (There exists, by the way, a special type of abstract—the "indicative" abstract—which provides little more than an "indication" of what the abstracted document is about, but that clearly has no place here.)

Relative weighting The designated objectives are usually best achieved with an abstract analogous in structure to the thesis itself, although in the interest of brevity you must severely limit what you include. For example, with an investigation oriented toward results, methodological considerations generally retreat into the background in the formal abstract. On the other hand, if techniques actually were the focus of your research, you might describe your experimental results only to the extent that they offer important validation of the methods. Note that it rarely proves appropriate to cite specific numerical values in the context of an abstract.

Passive imperfect   The distinction we suggest above between methods-oriented and results-oriented research is broadly applicable throughout the sciences, and it can even have an impact on linguistic aspects of an abstract. If observations, measurements, or other experimental results occupy the foreground in a project, then a summary of the questions that guided the work commonly takes heavy advantage of passive imperfect constructions:

Ex 9–1   The objective was to determine whether ...
(how much ..., under what circumstances ...)

which might continue with something like

Ex 9–2   For this purpose, ... was ...
It was assumed that ...
Measurements at XXX led to ...

Present tense   On the other hand, *answers* elicited with respect to the initial questions are more often best expressed in the present tense; e.g.:

Ex 9–3   This shows unambiguously that ...
The results demonstrate the advantages of...
By analogy it is clear that ... (... reasonable to exclude ...)
The temperature dependence of both the tensile strength and the 0.2%
elongation limit are free of yield stress anomalies at 800 °C.

It is not unusual to find in such a formulation at least part of the original question repeated verbatim.

In a more methods-oriented piece of work, however, it is a particular set of techniques that mainly needs to be addressed; e.g.:

Ex 9–4   ... a computerized system for continuous measurement of the oxygen consumption in ... was developed ...

One would proceed to discuss such things as what the method of interest has been shown to accomplish, how a particular apparatus functions and for what it might be used, and what advantages this specific procedure offers relative to other approaches. Especially welcome even in the limited context of an abstract are positive statements clearly demonstrating in what sense the method at hand is truly new, for example, or exceptionally reliable, or unusually precise. Examples of potentially effective phrases in this context would include

Ex 9–5   ... was perfected ...
... was successfully implemented under extreme X conditions ...
... produced data with a precision of YYY ...

This is another situation in which use of the present tense is commonly justified; e.g.:

Ex 9–6   ... the instrument array consists of ...
             ... additional advantages include ...
             ... can be simplified (improved) ...

Active descriptions   An especially desirable linguistic strategy casts the system under investigation in an *active* role:

Ex 9–7   ... transforms itself ...   ... increases with ...
             ... shows significant ...   ... displays no ...

As previously noted, one's abstract must always be so formulated that the reader will be able to understand it without consulting other parts of the associated thesis. It should also be kept entirely free of tables, figures, or other graphic elements, and must not make direct reference to specific features within the master document (tables, figures, formulas). The latter injunctions suggest some of the ways in which an "abstract" differs from the "synopses" that often accompany journal articles.

In the event you have employed numerical designations for things like chemical structures in your thesis, it can be helpful in the abstract to cite the corresponding number in conjunction with a verbal reference to the item in question (after it, in parentheses), thereby providing the reader with a valuable link between the abstract and the complete document.

Literature references   References in an abstract to the published literature always take a very generalized form, as for example

Ex 9–8   ... demonstrates for the first time that ...
             ... in contrast to reports in the literature ...

If your thesis happens to contain an especially comprehensive overview of the related background literature—one not readily accessible elsewhere—you may wish to indicate that fact in the abstract with a sentence like

Ex 9–9   Included is a literature review encompassing all previously reported information on this topic.

Extent of the abstract   In the absence of specific instructions to the contrary, the abstract for a graduate thesis should be restricted to a single page, consisting of no more than about 300 words. An undergraduate thesis abstract would of course ordinarily be much shorter. (The term "word" is here to be regarded as including abbreviations, numbers, and even unit symbols like "g" for "gram".) Despite this limitation, any reader—your advisor, a member of your thesis committee, or simply someone curious to know what you have done—must still be able to gain from a quick look at the abstract an accurate impression of the scope and impact of your work.

Type size   Quite often, thesis abstracts are set in type smaller than that used for body text (though—as usual—applicable university regulations must be respected). Smaller type, especially when coupled with narrower line spacing, greatly facilitates confining the abstract to one page.

Placement   The abstract of a thesis is typically to be found near the front: immediately after a preface or acknowledgements page, but ahead of the table of contents, analogous to the practice with most journal articles. Be aware, however, that some institutional guidelines specify positioning the abstract at or near the *end* of a thesis (as in Ex 8–7), and relevant university dictates must again be followed to the letter. Just as with a title page, dedication, preface, or set of acknowledgements, no explicit page number should appear on an abstract page.

The word "Abstract" itself typically functions as a title for the text of an abstract, set above the text and centered horizontally. It is often recommended that the rather uninformative word "Abstract" be preceded, however, by a restatement of the full title of the work, a sensible suggestion in many ways since it ensures that, in the absence of the complete thesis, the abstract will still be meaningful.

The following example will serve to illustrate a rather brief thesis abstract.

Ex 9–10   **Abstract**

The preparation, physical properties, and spectral characteristics of N-mesylhydroxylamine, $CH_3SO_2N(H)OH$ (**1**), are here reported and discussed for the first time.

Alkaline hydrolysis of **1** ($pK_a = 9.25$) is shown to proceed via a "nitroxyl" (HNO) intermediate in a process that can be represented as

$$\mathbf{1} \xrightarrow{\ +\,OH^-\ } CH_3SO_2^- + HNO$$

$$2\,HNO \xrightarrow{\ -\,H_2O\ } N_2O$$

(Although we offer the above as an example, it could be argued that this particular abstract is hardly exemplary from a documentation standpoint, in that the reaction arrows and the multilevel print demanded by the equations would effectively rule out certain forms of reproduction and dissemination.)

C 9–1 Locate formal errors in the following proposed abstract for a masters thesis, and suggest ways to improve it.

Abstract

The work described here was carried out in the Department of Chemistry between December 2001 and May 2002.

An attempt was made to clarify whether or not compounds of the type XY–R react with Z even when the residue R is electronegative, as in cases like R = A and B. Reaction does indeed occur at 80 °C with all the examples studied, provided a suitable solvent like S1 or S2 is present.

The resulting products, of the type Z–R, form according to the equation

$$XY\text{–}R + Z \longrightarrow Z\text{–}R + XY$$

(cf. for example Table 3–4 in Section 3.5). Such materials are thus now accessible in good yield (A: 90%; B: 95%) by a much more convenient route than the procedure described by Meyer and Jones [34].

C 9–2 In a masters thesis with the title

Removal of ZZZ from Exhaust Gases:
Optimization of a Filter Unit Containing Y-Silicate

the following significant results are reported:

– A filtering device, designed to contain only charcoal, was supplemented with a layer of a porous Y-silicate, a material not yet available commercially;
– It was discovered that, before use, the Y-silicate required tempering at 500 °C for ca. 30 minutes;
– Under normal operating conditions this system successfully removed 90% of the ZZZ present (charcoal: 75%);
– Unlike charcoal, residual Y-silicate need not be classified as a "hazardous waste" due to the extremely low solubility of ZZZ once it has been absorbed.

Try to compose an appropriate abstract for this report.

C 9–3 Critique the following abstracts from the perspectives of expressiveness, conciseness, etc. (The first example is derived from a journal article rather than a thesis, but the basic requirements are the same.)

a This paper describes both a specific efficiency analysis and an overall program of functional analysis carried out (as previously reported [12]) at the XXX Corporation, together with the chief conclusions drawn. The main goal of the program was to establish a well-defined and practical approach to improving product quality, plant productivity, and overall economic viability of the process. Also covered are the results of a supplementary computer analysis undertaken with the aim of increasing the applicability of the proposed scheme. The methodology applied in implementing the novel computer-based material-equilibrium approach (MASSBAL) to evaluating the program, as well as analyzing observed incremental improvements in the key individual production zones, is described here as well.

b X-Ray analysis of lead subjected to pressure has shown that at room temperature and a pressure of (130 ± 10) kbar the initially face-centered cubic structure is gradually transformed into a hexagonal closest-packing of spheres. The accompanying change in volume was found to be (-0.18 ± 0.06) cm^3/mol.

10 Introduction, Definition of the Problem

- Here we describe the purpose and chief characteristics of the "Introduction" to a thesis.

- After completing this unit you should feel confident about exploring the nature and goal of your research against a backdrop of the published literature, and be prepared as well to stimulate interest in the topic.

Q 10–1 What must one include in the "Introduction" or "Presentation of the Problem" section of a thesis?

Q 10–2 How is the Introduction dependent upon your familiarity with the literatur?

Q 10–3 Might this section also be entitled something like "Background Information", "Previous Work", or "Literature Review"?

Q 10–4 What should the reader learn by perusing your Introduction? What exactly should you hope to accomplish here?

Q 10–5 Is it the role of this section to reveal *your* understanding of the problem at the outset?

Q 10–6 What does the introduction to a thesis have in common with a review article in the published literature?

 | 2.2.6

Introduction, entry point, presentation of the problem

The "Introduction" to a thesis has the assignment of leading the reader into the subject matter that represents the essence of a research project. Since every piece of experimental work in the natural sciences is concerned with one or more questions directed toward nature, the reader will expect you to articulate in the Introduction the specific questions that served as the basis for *your* work. Some advisors even suggest use of the alternative designation "Statement (or Presentation) of the Problem" for this section, since it expresses more directly the real objective, and is less likely to encourage excessive historical digression.

In the Introduction you will want to describe the way you interpreted the fundamental challenge: *after* (!) you had familiarized yourself thoroughly with the relevant background material. This means you will need to devote some attention to questions like

– How and when did this topic initially take shape?
– What was already known when you began work?
– At that point, what precisely was it that was still unclear?

The Introduction is of course also the proper place to address anything you think would assist the reader in understanding your goals, while at the same time kindling an interest in the problem.

State of knowledge
Concentrate above all on the way you establish an overview of previous results in the area under investigation. Remember that you are trying to spell out as clearly as you can the general state of understanding when you first took up the problem.

Literature survey
No other part of a thesis or dissertation is as intimately connected with the literature as the Introduction, as a result of which at least a significant portion of it could legitimately be described as a "literature survey". It should thus have much in common with the review articles featured in many journals. Looked at from a somewhat different perspective, the Introduction can be interpreted as a picture frame within which your findings are to be put on display.

The source of ideas
In this section it is especially important that you pay attention to distinguishing between your own thoughts and discoveries, on the one hand, and those of others—an obligation that will become even more urgent in the subsequent "Results" and "Discussion" sections (see Units 11 and 12).

Problem definition
Problem definition in the context of scientific research is expressed as a combination of what is already known in a particular area together with certain things that are *not* yet known. A major investigation like the one expected to culminate in a doctoral dissertation may in the end actually address multiple problems. Even people with no direct involvement in science realize that a new scientific discovery often uncovers more questions than answers, and newly disclosed facts nearly always increase our awareness of how much actually remains unknown. It is with thoughts like these in mind that you should set out to structure your Introduction. In the process, by all means feel free to engender at least a hint of suspense.

Near the end of an Introduction there will often appear a sentence like one of the following, clearly reminiscent of the document's Abstract:

Ex 10–1   … The goal of the investigation thus became trying to find out if …
… For this reason it appeared reasonable to attempt …
… but there would obviously be no definitive way to proceed along these lines
without first determining …
… It therefore became necessary to clarify whether …

As the preceding examples suggest, it is entirely appropriate to include in
your Introduction a direct statement of the scientific problem with which
you were confronted at the outset. On the other hand, opinions differ on
whether an introduction should also disclose the outcome of the work. If
so, it will of course become much more difficult for you to maintain any
air of suspense in the Discussion section that follows, but no one really
expects your thesis to resemble a detective story, in which the culprit may
not be unmasked until the last page. Especially with a complex investigation it can in fact be to your advantage to announce the decisive new discoveries immediately, in the Introduction:

Ex 10–2   … whether X or Y is more applicable. The results presented here suggest
that … is compatible only with X.
… through which it became possible for the first time to establish … in an
unambiguous way.

We now turn to ways in which the Introduction might be structured. Consider, for example, the outline in Ex 10–3.

Ex 10–3   1   Introduction
1.1 History of the Application of Method X
1.2 Survey of the Chemical Properties of Compounds of Type Y
1.3 A New Approach to Bicyclic Y Systems?

In the first two subsections the author is clearly setting the stage for a problem definition, which then appears in subsection 1.3, approached in the
form of a question. (In essence, "I was hoping to synthesize compound Y
using the X approach. Did I succeed?")

The opening sections in the next sample outline, Ex 10–4, suggest careful
analysis and characterization of a specific problem, leading to a particular
novel form of solution devised by the author. Societal ramifications
apparently play an important role here, and these are examined not only
from a scientific perspective, but a legal standpoint as well. The seemingly
benign heading of Section 1.4 presumably heralds this author's unique contribution to dealing with the broader dilemma.

Ex 10–4   1   Introduction and Problem Definition
1.1 The Use and Ecological Impact of Halones
1.2 Legal Considerations
1.3 Standard Approaches to Halone Disposal
1.4 Modification of Halones by Catalytic Hydrogenation
2   Results and Discussion

The Introduction constitutes the first element in the main body of any thesis. Sometimes it is nonetheless treated as a separate, independent "Part", which is then integrated into the whole, as in the following example:

Ex 10–5 Part I   Introduction
1   XXX
2   XXX
Part II   Experimental
3   XXX
4   XXX
5   XXX
Part III   Results and Discussion
6   XXX
7   XXX
Bibliography
Appendix

Chapters 3 and 6 would function here as "mini-introductions" to Parts II and III.

The Introduction is often a logical place to underscore, among other things, methodology you have elected to apply in your attempt to solve the overall problem. Phrases pointing in this direction could take forms like

Ex 10–6 … With this objective in mind, …
… In order to distinguish between …
… Since it was important to prevent …
… For this reason it seemed prudent to undertake a new study of XX
  with special attention to …

"Materials" Apart from historical information and methodology, an Introduction can also be used to address at least briefly the subject of "materials". For example, in a biological study, a little attention might be directed here to such things as the species investigated, sampling techniques adopted, and statistical evaluation of the experimental results. All these become subject to detailed consideration in later sections of the thesis, but it is still important at the outset to ensure that the reader will be in a position to follow (sympathetically!) those subsequent treatments.

General guidelines There are no generally recognized standards or guidelines—regarding length, for example—applicable to the introductory section of a thesis. This is in contrast to the abstract and the title, which must be made suitable for direct inclusion in established public information sources and documentation systems. You nevertheless need to be sure (as usual) that your university has not come up with its own set of rules that must be respected.

Extent Sometimes it is possible to craft a satisfactory introduction that consumes as little as a single page (cf. C 10–3). If you exceed about five pages the reader might become suspicious that you were more concerned with fill-ing space than carefully erecting a user-friendly portal into a specialized body of information. In general, we feel roughly ten pages should be considered as an upper limit for the Introduction to a thesis based largely on experimental data; as advisors we would in fact be inclined at that point to draw a rather firm line.

First person constructions

It has become increasingly common in published scientific works to encounter—alongside the impersonal, passive sentence constructions that dominate not only our illustrations but also most of the scientific literature—examples of first-person verb forms, which unquestionably add life to their surroundings. Note that this development carries an important implication, however: the authors responsible are suggesting (or admitting) in a subtle way (perhaps unconsciously!) that the parties involved in conducting a scholarly investigation may themselves play a more than passive role:

Ex 10–7

... This being the case, we elected to ...
... We therefore separated (introduced, heated) ...
... In the hope of conferring antimalarial activity on pharmaceutical precursors, we prepared ...

The "I" form

What at first glance appears to be a "stylistic" matter thus constitutes in fact a substantial break with a tradition that—at least over the last century or so—has been maintained rather rigorously in nearly all scientific communication: formal anonymity. Apologists justify reliance exclusively on passive expressions on the grounds that "objective" scientific results should not have the appearance of being subject to influence from the persona of the practitioner(s). Participants must disappear into the background! Others argue vociferously that, irrespective of the merits of this concern, active-voice sentences are highly desirable, not only because they are much simpler to construct and also comprehend, but they are considerably less likely as well to be mutilated by grammatical errors. Some would go further and contend that passive constructions are actually harmful, in that they tend to *over-exaggerate* the "pure objectivity" of science. We leave it to you to decide whether or not to welcome the pronoun "I" and its kin into your thesis. (For our part, we would still not be so inclined, most especially in the Introduction and Results sections.)

It can be a valuable exercise to examine introductory portions of an assortment of journal articles as potential models for your thesis Introduction. In the process you will encounter authors who succeed in accomplishing all that needs to be done with just one or two finely tuned sentences—in many of the articles in the journal *Angewandte Chemie International Edition*, for example. Journal articles are of course often derived from doctoral dissertations, an observation that underscores their relevance in the present context. The following illustrations—to which we have appended comments—may help reinforce the point.

Ex 10–8 Although compounds with direct X–Y bonds have for some time been employed in organic synthesis [1], their structures were first established with certainty in the 1980s [2–12]. Our own interest in X–Y compounds was aroused by isolation of Z [10] ...

[An extensive synthetic history has here been adroitly dispatched with a single reference, [1], which turns out to be a 40-page annotated chapter in a handbook. The reader's attention is then directed toward one specific aspect of the historic record—structure elucidation—as documented in 11 original papers. The terse reference to the relatively recent appearance of these papers is probably intended to infect the reader with at least a little of the writer's own enthusiasm. (The article from which this excerpt is drawn of course contains element symbols rather than the generic "X" and "Y" shown here, and a chemical formula where we have substituted a "Z"). At the point where our fragment leaves off ("... isolation of Z [10] ...") the author had not quite finished presenting introductory material—which this particular journal, unlike some, does not set apart with special typography—but one can already deduce where he or she was headed: A new compound related in some way to "Z" has now been prepared, its structure has been confirmed, and it shows promise as a synthetic intermediate.]

Ex 10–9 The X–Y system is known [1, 2] to play a central role in the pathogenesis of hypertension, in part because of the presence in the circulatory system of vasoconstrictor Z [3]. Inhibitors of the enzyme responsible for converting Y have long been candidates for therapeutic evaluation [4–9]. Given the presumably greater selectivity of X-inhibitors [10], however, the latter should without doubt be investigated as well, ...

[The biochemical—and therapeutic!—significance of a complex physiological phenomenon has been efficiently summarized again, with just two introductory sentences. In conjunction with the title of the article, this fleeting glimpse is sufficient, however, to permit the interested reader to anticipate with some confidence what will follow, at least in a general way: an account of the preparation of a superior class of X-inhibitors. The reader's curiosity has also been effectively piqued: How and to what extent was the envisioned goal achieved? Are relevant pharmacological results already available? (No, as it happens, but in a concluding sentence—which we have omitted—the authors affirm that several promising compounds have indeed been prepared and carefully evaluated from the standpoint of their enzyme-inhibitory activity.)]

Hopefully these examples will convince you of the wisdom of studying the introductions to a number of recent publications before you embark upon preparation of the Introduction to your thesis. Some of the exercises ("challenges") that follow may further stimulate your creative impulses at this point.

C 10–1 Critique the following (extraordinarily concise) opening to a thesis:

Introduction

Our laboratory has long been engaged in studying the synthesis of X. The work reported here reveals that—contrary to the claim of Young and Smith [1]—compound X can be obtained quite readily from the industrial byproduct YY.

Results

...

C 10–2 Compress the following background information into a form suitable for an Introduction.

– Mayer, 1921: Discovery of Y, the first example of a compound of the class XXX; laboratory preparation of very small quantities

– Miller and Jorgenson, 1955: large-scale synthesis of Y facilitated by a new route via Z

– Chiang, 1958: Spectroscopic investigation of Y, and establishment of its fundamental chemistry

– Peters, 1961: Utilization of Y as a starting material for preparing pharmaceutical agents from the compound class DDD

– Miller, 1969: Y becomes commercially viable through control of the synthetic reaction such that very little (ca. 10%) of byproduct A is formed

– Goal of this investigation: Laboratory-scale variation of the reaction conditions (especially temperature and solvent) in the reaction used to prepare Y commercially in an attempt to minimize the formation of A

C 10–3 How do you react to the following introduction to a dissertation from the biomedical field? Can you suggest ways of improving it? Explain.

Introduction

Pharmacological treatment of patients with severe cardiac insufficiency has been shown to prolong survival. This is especially true with chronic congestive cardiomyopathy, in which administration of vasodilators is known to be extremely effective, as manifested in both objective and subjective increases in cardiac sufficiency.

Relatively little attention has so far been paid to rheological and hematological changes experienced by patients suffering from low-output syndrome. Other studies document hematorheological changes in conjunction with cardiovascular diseases like morbus XXX [1, 2, 3, 4, 5], but there is still no consensus regarding the clinical significance and possible therapeutic ramifications of these findings. Only with morbus YYY has it been demonstrated that impressive levels of clinical remission can be achieved through fibrinogen reduction based on plasmaphoresis [1]. With more severe arterial obstruction, fibrinogen can be reduced most conveniently by administering snake venom, which leads at least temporarily to improved circulation [6, 7].

It may be that hematorheological changes play a more important role in blood supply to tissues in patients with marginal cardiac function relative to healthy individuals.

The goal of this project was to examine blood rheological factors in such patients in the hope of establishing their relevance to cardiac function, microcirculation, and thrombo-embolic events.

11 Results

● This unit explores what should be incorporated into the "Results" section of a thesis, how boundaries for the section should be established, and what approach you should take in describing your results.

■ The suggestions here should help you express your findings in a way that will be both efficient and lucid.

Q 11–1 Must a thesis always contain a section labeled "Results"?

Q 11–2 How does the "Results" section differ from the""Discussion"? Under what circumstances might the two be combined?

Q 11–3 Which findings, measurements, etc. belong under "Results", and which in the "Experimental" section?

Q 11–4 Should results be described in the past or the present tense? Active or passive voice?

Q 11–5 Is this the place to examine procedures in detail?

Q 11–6 To what extent is the Results section an appropriate place to state conclusions?

2.2.7

The "standard" outline

We begin by considering the first of the questions posed above (Q 11–1), to which we respond with an emphatic–"no": it is certainly *not* essential that a thesis include a separate section entitled "Results"—unless of course you find yourself subject to a set of university requirements that state otherwise!—but we will nonetheless assume its presence (cf. the "Standard Outline" on p. 54). Such a section is almost always followed immediately by another called "Discussion" (cf. Unit 12). Although we have encountered many excellent theses that do not conform to the standard outline, the fact remains that most journals insist that articles submitted for possible publication be based on the conventional structure, and for this reason alone the novice anticipating his or her public debut in print is well advised to become familiar with traditional presentation style.

Presentation of your findings

A Results section serves to communicate important new discoveries relevant to the research question(s) posed initially. This generally marks the frontier of the "heart" of a thesis, the part devoted to the author's original contributions. (Occasionally there will be an "Experimental" section sandwiched between the Introduction and the Results, in which case that could serve as an alternative entryway into your personal efforts, albeit one rarely considered an attractive choice.) Whereas the Introduction represents a sort of intellectual forerunner, in the sense of providing preparatory grounding both in the problem at hand and its context, you are now invited to put on display your own proud accomplishments. For the moment you should not dwell on their significance, however: save that for "discussion" in the section to follow (see Unit 12).

Nature and scope of the work

You have already set the stage for your project in the Introduction, so there should be no need here to offer detailed justification of individual experiments or the relevance of your results. Announcing the salient findings will suffice. The reader is for now interested only in ascertaining the ultimate scope of your investigation, its concrete outcomes, and the level of care and originality you brought to the work.

"Results" vs. "Discussion"

Later, in the Discussion, readers will of course wish to learn more about how you think your results should be interpreted, and something of their likely importance in the light of what was previously known. Indeed, this structural distinction between "results" and "discussion", and the clarity it adds, is the main reason we recommend a division into separate sections. Mixing the two complicates not only the reading of the document, but also its preparation, because closer attention must then be paid to differentiating between fact and interpretation, and between your contributions and those of others.

Especially with a methodology-oriented project—development of a procedure, for example—experimental details should appear in the Results section only to the extent absolutely necessary to ensure comprehension. For example, if it would otherwise be difficult to see the point of some of your results you might need to explain briefly why you designed a particular set of experiments the way you did, or the role played by a certain piece of apparatus. Painstaking descriptions of individual components or materials you utilized, and of laboratory procedures or field conditions—these should be set aside for the Experimental section.

Reliability   On the other hand, you must *not* relegate to a subsequent formal description of procedures any factors crucial in assessing the reliability of your findings, or the true "robustness" of a heralded innovation, for example, especially given the likelihood that the average reader would almost certainly overlook this essential information if it were buried in the midst of a host of minutiae. Never allow the "fine print" (n.b.: small type is in fact *literally* utilized for presenting experimental details in many journals!) to become a repository for potential weak points in your argumentation.

The information content of phrases like

Ex 11–1   … recrystallized from C_2H_5OH/H_2O (4:1) …
… with the oval window open, the sample was …
… immersion fixation with phosphate-buffered 2% glutardialdehyde (pH 7.4) …

definitely *should*, however, be put off to the Experimental section. That said, establishing optimal placement for specific factoids can admittedly be one of the trickiest parts of crafting the formal presentation of one's efforts.

Graphic elements   The notion of "results" can be extended—at least in important cases—to encompass even graphic elements: a photomicrograph of a tissue sample, for instance, or an electrophoretogram of a homogenate, or an NMR spectrum of a synthetic intermediate. A structural formula or a reaction scheme would not class as a result, however, and definitely belongs in the Discussion, unless its presence in the Results section is absolutely central to ensuring comprehension.

Numerical results   Many results in the natural sciences as well as engineering and medicine take a quantitative form: data expressed as the combination of a numerical value with one or more units. Information of this sort is often communicated most effectively in a table or a graph. Graphic interpretation can be an especially powerful means of making numerical data both intelligible and appealing.

Results like these should never be dismissed as "simply numbers", by the way. Each deserves to be treated as the bearer of a tiny but unique message. One of your challenges is to make numerical data in the Results section of your thesis not only manifest, but also articulate. If you proclaim that a specific physiological variable

Ex 11–2 a   … displayed an average value of (78 ± 4) for 35 high-performance athletes, in contrast to (69 ± 3) for the 20 representatives of a control group …

you are in fact serving notice that you took the trouble to assemble a great many numerical measurements, compiled and examined them in a particular way, and then established both a magnitude and a "reliability" for their average. But expressed in the bland fashion proposed above this revelation could never claim a prize for "lively communication". It shows little evidence of being directed to a specific reader, for example, and it certainly does nothing to encourage the reader to pause and reflect. Consider the following alternative, however:

b … the median X-value derived from 35 high-performance athletes was significantly (ca. 11%) higher than the corresponding median from 20 representatives of a control group, namely (78 ± 4) mg/L vs. (69 ± 3) mg/L (standard deviation $\rho < 0.02$) …

Surely you will agree that the latter makes the point much more captivatingly! As someone anxious to disseminate what you believe to be significant news, you will of course want in the subsequent Discussion section to elaborate further upon this result, addressing the relevant circumstances, the importance of what you have learned, and its potential consequences, but for now the refined statement above at least ensures that the essence of your discovery will be accorded some attention.

Order of presentation

One of the greatest challenges in preparing a thesis manuscript is organizing the content as effectively as possible: In what sequence should you articulate your thoughts? We look now a bit more directly at what is entailed in devising a first-rate outline:

- Begin by clearly separating—through informative chapter or subsection headings—the various topics that would lend themselves well to independent consideration; e.g. problem area A from problem area B, synthetic issues from analytical results, data collection from calculation and interpretation;
- Try to arrange information of equivalent hierarchical rank in a consistent way; e.g., proceeding from complex to straightforward, from small to large, or from central to peripheral;
- Structure the material for maximum impact as opposed to providing a strictly chronological record;
- Keep in mind that you will later want to arrange your Experimental section so that it parallels your Results;
- Establish the objective of "telling a story", with a beginning, a gradual unfolding, and a "happy end", and encourage each thread to evolve naturally from the one preceding it; make sure, incidentally, that episodes related to sideshows (because they have little to do with the central message) are readily recognizable as such—or else omitted altogether.

Literature references

Published literature will contribute almost nothing to your Results section. Here your mission is to deal as single-mindedly as you can with your own original work. Formal reference to the work of others might conceivably be appropriate with respect to a methodological comment, or perhaps in the context of a comparison, but it is generally better to postpone acknowledging the developer of a technique, say, until the Experimental section.

A rule of thumb many journalists recommend is worth citing here with regard to the organization of individual subsections (containing multiple paragraphs), and even for paragraphs (constructed from several sentences): "Guide the reader along a path that proceeds from most important to least important"; e.g.,

Ex 11–3   The ^{1}H-NMR spectra ($CDCl_3$) of both **7a** and **7b** feature prominent singlets—attributable in each case to a pair of olefinic protons—at 6.43 and 6.20 ppm, respectively. The observed downfield shift points to **7a** as the α-isomer, since this phenomenon is indicative of interaction between the ethylenic moiety and an α-NH group.[3] A similar pattern is observed for the two isomers **X** and **Y**. Dipolar solvents tend to interfere with intramolecular interactions of this type, as illustrated by the fact that in d_6-DMSO the olefinic signal for **7a** appears at 6.27 ppm, whereas the spectrum of **7b** remains virtually unchanged (6.24 ppm).

Here the most important information was incorporated into the first two sentences: certain experimental parameters were measured and interpreted, which in turn made it possible to assign structures. The second sentence, with its passing reference to the literature, contains no true experimental results, but is nevertheless essential in associating a particular structure with compound **7a**. The last sentence's terminal placement suggests that from the author's perspective the dipolar-solvent effect is of secondary concern.

To avoid seriously disrupting the flow of ideas, large quantities of data should be reserved for an appendix, with only a summary in the Results section, accompanied by at most a few representative examples. For instance,

Ex 11–4   ... At $\vartheta = 40\ °C$ the concentration of S, R = Me, drops off rapidly, and after 24 min has fallen to 0.015 mol/L (see also the middle curve in Fig. 3). Additional data documenting changes as a function of time—including at higher temperature, and with X = H, Et, iPr—are provided in Appendix A.3 ...

On the other hand, be sure to make explicit reference in the Results to any of your findings that are unusual, and to extreme cases, because these will necessarily be subject to later discussion; e.g.,

Ex 11–5   ... by contrast, the residual solution turned violet under XXX conditions ...
... the YY content in blood exceeded a value of ZZ only at ...
... the impurity level reached a minimum when filter F1 (Fig. 40) was packed with activated charcoal. Lime, on the other hand, proved to have a low retention capacity ...

Results that are in some way associated, analogous, or otherwise subject to comparison are generally best expressed in tabular form (see Unit 20); e.g.,

Ex 11–6 … The outcome varies depending upon whether the packing agent is XX or YY, as shown in Table 5–7 …

Functional relationships, by contrast, lend themselves better to presentation as diagrams or graphs (see Unit 21.3), preferably ones that also indicate error limits. Again, postpone all descriptive comment until the Discussion, including interpretation of the path followed by a particular curve. Some advisors insist that theses contain full, explicit disclosure of all experimentally determined values, if only in an appendix, although details at this level must usually be dispensed with in published papers.

As already suggested, you are expected to save your concern with implications and the potential significance of a set of results for the Discussion section. Even so, it is not always easy to decide where to say what. It may help to consider your options from the point of view of a distant reader—not your advisor or another member of your thesis committee, but someone joining the research group in the future, perhaps, or an outsider in search of specific information.

The Results portion of a thesis usually benefits from being broken down into several subsections. In the case of a chemical investigation, for example, a scheme like the following might be appropriate:

Ex 11–7 …
2   Results
2.1   Preparation of Starting Materials
2.1.1   Substance X
            Synthesis
            Characterization
2.1.2   Substance Y
…

[We have left unnumbered here the segments devoted to synthesis and characterization of X in order to avoid 4-place identifiers (2.1.1.1, 2.1.1.2), consistent with the advice in Unit 8.]

You should of course have a rational basis for whatever framework you devise: underscoring the original set of goals, for example, or possibly reflecting more clearly your tangible accomplishments.

Interim reports Actually, this part of your thesis ought to be relatively easy for you to compose, especially if you were conscientious during your research about regularly preparing interim reports (see Units 1 and 3). In the ideal case, much of the Results section (just like the Experimental section) will almost write itself as you "cut and paste" appropriate passages from your earlier work.

Presentation style Results tend to be of two fundamental types, and the distinction is one that should have a subtle influence on your writing style. Some reported findings will essentially echo descriptive passages in your notebook, and these are best subjected to passive treatment, as in Ex 11–2 a, b. At the same time, as the author you will be anxious to ensure that certain key one-time events (in the past) are elevated to the status of lasting assertions, and these warrant expression in the present tense, as with Ex 11–3. The following rather typical sentence fragments underscore the point:

Ex 11–8   ... therefore differ significantly in thickness.
             ... an important consequence of this interpretation is ...
             ... takes an alternative course.

Another appropriate occasion for use of the present tense is when you wish to summarize a set of findings, or cite specific data points drawn from a figure or a table; e.g.,

Ex 11–9   ... The path of the resulting curve is similar to ...
             ... is suggested by the anomalies apparent at ...
             ... displays a pronounced maximum at ...

This also applies to declarations of the form (cf. Ex 11–6)

Ex 11–10   ... can be simulated by Equation (3) ...
             ... is shown in Fig. Y ...

We are dealing here not with specific events or characteristics that should be regarded somehow as time-dependent, but rather with ongoing conditions or phenomena.

Finally, make your objectives clear to the reader throughout your exposition. Don't be one of those writers who seem intent on spewing forth a relentless torrent of isolated facts. Maintain a consistent focus on your goals, and refuse to settle for anemic pronouncements like

Ex 11–11 a ... X was repeated in the course of a second series of experiments under Y conditions ...

when minor rephrasing would reward your reader with the much more pointed

b   ... In order to ascertain whether (rule out the possibility that, assure that, ...)
             ..., X was repeated under Y conditions.

◇

C 11–1   What argument can you make for treating results and their discussion separately in a thesis?

C 11–2   Which of the following elements of a thesis appear to be strong candidates for inclusion in a Results section?

Infrared spectrum of a compound, photo of a skin lesion, recorded pH value, calculated voltage, findings borrowed from another research group, drawings of several pieces of apparatus, graphical presentation of a set of data points, interpretive comments regarding a particular instrument reading.

C 11–3   Suggest informative subsection headings that could be applicable to a few of your own findings, and thus potential candidates for the Results part of your thesis.

C 11–4   When is it appropriate to use the present tense in conjunction with information in a Results section, and when the past?

C 11–5   Identify portions of the following excerpt that seem most applicable to thesis sections labeled (a) Results, (b) Discussion, and (c) Experimental.

As a way of following the reaction, the UV absorption [λ_{max} = 280 nm, ε = 0.0045 L/(mol·cm)] as a function of time was measured for a solution of 31.5 mg of **2** in 5 ml of acetone. Compound **1a** was obtained in high yield (> 90%) when the reaction was carried out in acetone or ether. In trichloromethane or carbon tetrachloride, however, the yield of **1a** never exceeded 20%, with **1b** instead being the major product. This confirms findings of HARTZFELD et al. [13], who prepared **3a** and **3b** in nonpolar solvents, where R = ZZZ ...

C 11–6   What organizational principle has apparently been applied to the following text? Would you regard the passage as informative?

In X-screening with mice, anilides **10** and **13** caused visible central nervous system effects at a dosage of 200 mg/kg. The former acted more rapidly than the latter. An LD$_{50}$ value of 64 mg/kg could be confirmed in each case on the basis of a Y test.[8] Derivatives **12** (both **a** and **b**) not only showed serotonin-antagonistic potential, but also led to a reduction in blood pressure, probably as a result of interaction with the Z system.

12 Discussion

- Immediately following the "Results" we come to what is arguably the most important part of a thesis: the "Discussion".

- In working through the suggestions below you will come to appreciate what actually belongs in a thesis Discussion (and what doesn't!), and how this section should be structured.

Q 12–1 What is the main difference between the Discussion and Results sections of a thesis?

Q 12–2 Might one not simply combine the Discussion and Results?

Q 12–3 Is it common to find conclusions and suggestions for further work in the Discussion section?

Q 12–4 Must every thesis contain a section entitled "Discussion"?

Q 12–5 What is most appropriate with respect to voice and tense in the Discussion part of a thesis?

2.2.8

The "essence" of a thesis

The "Discussion" might be described as the "nerve center" of a thesis. In combination with Introduction, Results, and Experimental sections this represents the essence of the document as a whole. It is here in the Discussion that the reader will most clearly acquire a sense of how well you as an author are able to think in contextual terms and develop a persuasive argument.

The overall goal

A good thesis Discussion will necessarily be evaluative in character. That is, your newly acquired research results (cf. Unit 11) must now be subjected to rigorous analysis and then closely examined in the context of what was previously known in the field. To put it yet another way, your personal findings are here to be merged with information from the past in a bold attempt to define and articulate a new body of public knowledge.

Contributions of others

Needless to say, your writing must be absolutely clear with respect to what it is you yourself have contributed to the subject. This requires labeling all facts or ideas attributable to others so explicitly—with rigorous citation of appropriate sources of further information—that everything else in the section can by definition safely be regarded as yours. For example, the following passage

Ex 12–1   ... based on reported IR [7–13], Raman [8, 12, 14], and NMR [15–17] studies it is apparent that an earlier structural assignment [3] is ...

has been scrupulously annotated, and one would quickly conclude that it contains no information stemming from the author's personal efforts (assuming none of the citations present are to documents he or she has published previously, which rarely turns out to be the case in a thesis). Instead, the comment is presumably serving as a platform from which to address new findings that do arise from the present research. On the other hand, the writer's own achievements *are* clearly central to

Ex 12–2   ... the broad absorption band at 1640 cm^{-1} cannot be attributed either to a C=O stretching vibration from X(H)C=O (X = CH$_3$: 1225 cm^{-1} [21]; X = C$_6$H$_5$: 1234 cm^{-1}, personal observation) or to skeletal vibration, which for **2** is shown instead to be at 1350 cm^{-1} (see Fig. 3–8) ...

or

Ex 12–3   ... When XXX functions as a reaction partner, the regioselectivity of the process—contrary to the assertion of Schill and Gorz—is drastically altered (cf. also the second column in Table 5) ...
... if ... then the voltage at node 5 (see Fig. 6) is shown to rise to 1.54 mV.
... all three varieties studied reveal high neuron density in their upper and middle regions (see Fig. 17 a–c).

Inappropriate omission of a relevant literature citation might easily lead to a suspicion that you are trying improperly to assume credit for someone else's work. Warding off all such doubts is the motivation behind explicit formulations like

Ex 12–4a   ... contrary to what they postulated,[14] the present work indicates that ...
... On the other hand, from results reported here it is clear that ...

or, even more pointedly (elimination of a "that" clause; cf. Unit 4),

b   ... as these results, in contrast to an earlier suggestion,[14] unambiguously confirm.
... as has now for the first time been dramatically established.

The various research findings you wish here to discuss need to be introduced very directly. This means you will inevitably be forced to tolerate a certain amount of redundancy from the standpoint of the overall thesis. It would be a mistake, for example, to assume that your reader is in full command of everything presented earlier—such as in the Results section. It may well be that he or she will not even have *read* the passage in question, despite its prior placement in the thesis. You must go out of your way to make certain that all ensuing arguments—taken by themselves—will be easy for a reader to follow.

Obviously you are anxious to spotlight positive aspects of your work, but it is proper also to devote attention to things you did *not* succeed in accomplishing, and to draw comparisons between the current overall situation and the one you envisioned when you first decided upon your goals.

Logic rather than chronology
: As you organize your thoughts, make a conscious effort to distance yourself from the actual chronology of the research experience. Experimental work is by definition a journey into the unknown, fraught with detours and dead ends, but all such obstacles are basically irrelevant from a scientific standpoint. Look upon your dissertation—as you would any other research report—not as a memoir documenting and explaining your every activity, but instead as a proclamation of a set of new insights, and as the public unveiling of a personal "thesis", in the original sense of the word (i.e., a postulate, or a proposition, advanced as part of an argument). How you achieved your results may perhaps play some role in your mentor's appraisal of your efforts, but otherwise it will be of interest only to the extent that certain intimate details may be critical from a methodological standpoint.

Your "story"
: You have already "introduced" the main part of your thesis through an Introduction (cf. Unit 10), in which you posed the formal question(s) to be regarded as the starting point for your odyssey. What you discovered in the course of trying to answer the question(s) has already been revealed as well—in the Results section. How you *interpret* your findings, and how they should be viewed in the light of what was previously known—in other words, your "answer(s)" to the original question(s), in the broadest sense—is what the engaged reader will be seeking to learn in the "discussion" that represents the story's conclusion.

The beginning
: Your Discussion should open with a consideration of those parts of your research that are of greatest interest. Start in the very first sentence with the most important message you wish to convey. (Newspaper editors also recommend this approach, by the way: Immediately below a headline they expect an article to commence with a pithy declaration, on the basis of which a reader can immediately judge the extent to which that article is likely to be worth reading.) Thus, your first sentence might well encompass a concise response to the definitive question: Did your original idea work, or didn't it? Does the postulated phenomenon occur or does it not?

In framing this central revelation, feel free to borrow appropriate words or phrases directly from the Introduction; e.g.:

Ex 12–5a (Introduction:) For this reason it appeared important to determine whether or not one can rely on chromatography as a reliable way of establishing a set of enantiomeric ratios for the constituents of peppermint oil.

b (Discussion:) Chromatography with a cyclodextrin-coated capillary column has been shown to be a reliable method for separating the constituents of peppermint oil, thereby giving rise to reasonably accurate estimates of enantiomeric ratios.

One answer, many answers
A sweeping lead-off assertion like the preceding one must of course be "unpacked" in the course of the sentences and paragraphs that follow. This particular author might feel obliged, for example, to explain in detail—with the help of carefully crafted tables and diagrams—that the separation in question is dependent both upon the use of very fine capillaries (inner diameter 0.18 mm) and also rigorous temperature control, and that despite experiments with a wide range of stationary phases, only cyclodextrin was found to be effective. The (presumably differing) separation efficiencies observed with respect to individual components of the mixture (menthone, isomenthone, menthol, etc.) would also need to be detailed, with perhaps an attempt to interpret them.

Elaboration
Once the major issues have been dealt with, our author would probably want to describe and elaborate upon some of the specifics of his or her findings, such as the precise influence of temperature, and problems encountered during separation attempts based on alternative coating agents, or at exceptionally high column temperatures. The impulse to digress a bit is actually to be encouraged so long as it doesn't evolve, say, into a new theory of transition states in adsorption kinetics—in other words wander too far afield.

Methodology
It is often appropriate in the Discussion part of a thesis to analyze *methods* separately from the results themselves, with both parts further subdivided as appropriate. In a project with a strong methodological component it would be important, for instance, to compare the chosen approach—its strengths and its limitations—with alternatives already familiar from the literature. What potential advantages were you attempting to exploit? In what sense do you see your method as being simpler, faster, more reliable, more broadly applicable, or more specific? Is it less demanding from a materials standpoint? Or more practical largely due to the tiny samples at your disposal? Possibly you found it less susceptible to interference. These and related criteria could all be worth touching upon as the Discussion unfolds.

The disadvantages of an experimental method warrant mention as well. The true value of a process can in most cases be properly appreciated only in the light of its inherent limitations. Indeed, the primary objective in some thesis projects is elucidation of just such shortcomings.

On the other hand, breaking ground with a *new* procedure may not have been a priority for you at all, but rather exploring the utility of an already widely adopted approach under a novel set of circumstances: "Would method X also be effective in situation Y?" Perhaps your investigation should be regarded as a kind of pilot project, with the goal of finding out if or within what bounds a certain methodology might be worth incorporating into your particular work environment.

Statistical backing Scientific research of almost every type is likely to benefit from some attention to statistical analysis. Usually this arises in the context of attempts at further buttressing one's conclusions. Note that addressing internal validation of your work may require you to defend the logic you applied in selecting a set of test cases, or explaining why you were satisfied with a confidence level of, say, 95%.

Range of applicability Any discussion of the relative merits of various analytical methods should take into consideration threshold levels, as well as vulnerability to influence from extraneous factors. Put somewhat differently, make a point of addressing any constraints you are aware of regarding the extent to which assertions you make are valid. Thoughts along these lines were what led us in conjunction with the problem underlying Ex 12–5 above to raise the matter of temperature restrictions, and to underscore the fact that of all the stationary phases investigated, only one proved effective.

Breadth of validity Even if your particular work is oriented more toward results than methods—e.g., "Does A react with B?", "Can Y exert an influence on X?"—you will still need to devote attention to evaluation. How trustworthy are the observations you report, for instance? In this context, pay special attention to assumptions and boundary conditions, and delineate clearly what you have *not* investigated, and thus situations about which you are in no position to make meaningful predictions. In this regard a thesis is quite different from a patent application, for example. Claims made in a thesis (or a journal article) must be restricted exclusively to what is *known* to be valid, whereas patent offices are notorious for their rather more generous attitude along those lines.

Figures and tables Tangible accomplishments often lend themselves to effective summary presentation through the media of tables and figures (cf. Units 20 and 21). To the extent that you take advantage of this avenue, be sure you refer explicitly in your Discussion to all such elements you include, even if this requires a certain amount of backtracking to the Results section. If, for example, the Results contains statements of the type

Ex 12–6a   ... the outcome of this series of experiments is presented in Table X.
... follow the unusual pathway depicted in Fig. Y.
... is linear throughout the range $2 \cdot 10^{-4}$ mol/L to $8 \cdot 10^{-2}$ mol/L (Fig. Z).

the referenced figures and tables should be subjected to a certain amount of comment and interpretation in the Discussion, as in

b   ... the shoulder at 270 nm in Fig. 2 (curve 1) can be attributed to contaminant A, present to the extent of ca. 5%; cf. the gas chromatographic trace in Fig. 3.

"Professionalism"   In discussing figures, be respectful of your reader's professional sensibilities. For example, when commenting on data displayed graphically, make use of appropriate technical terminology (maxima, minima, inflection points, etc.) if you wish to call attention to specific features you think important. The Discussion in a thesis should certainly reflect a level of scholarly discourse consistent with an informed audience.

Table guides   In the case of tabular presentations, actively assist the reader by pointing out what you consider to be noteworthy factors. In other words, feel free to augment the "legibility" of your tables with helpful remarks like

Ex 12–7   ... fails to reveal any correlation between values in columns 2 and 7 of Table X.
... whose uniqueness is especially striking (Table Y, line 3).

Hypotheses   Be on the alert generally for places where a curious reader would welcome a certain amount of interpretive elaboration. What has caused the reported behavior? How did it come to pass? Often the answer to such a question will depend upon an assumed mechanism or model, from which the phenomenon of interest then becomes predictable, but that type of explanation usually entails hypothetical elements. The validity of a postulated mechanism would rarely qualify as a certainty. In the event that multiple explanations might be advanced, present the alternatives of which you are aware, and weigh them as best you can. If several models appear consistent with your results, treat all of them objectively before explaining why you tend to favor one over the others. If another research team has offered a different explanation from yours for a phenomenon you are investigating, defend the stance you are taking as dispassionately as you can.

Apparent discrepancies   Suppose you have results that are in apparent conflict with a published report. Confront the discrepancy head-on, but avoid projecting the image of an all-knowing judge. Instead, calmly explore the opposing points of view, holding one outcome up against the other and perhaps allowing the reader to draw appropriate conclusions. The seeming disagreement might actually be found to disappear upon closer investigation—more careful selection of the test cases, for example, or unrecognized differences in experimental conditions—when additional factors have been taken into account. In any case, arrogance is never an endearing characteristic, and its intrusion is unwelcome especially in the context of a thesis or dissertation.

Future work   Near the end of your discussion you may want to be bold and suggest interesting avenues for future exploration: promising approaches to resolving remaining uncertainties, ideas for broadening the scope of a methodology you have developed, or ways of possibly refining your results. If so, be sure the suggestions you make are concrete. In place of a vague—and essentially trivial—observation like

Ex 12–8a   ... warrants further investigation.

supply something tangible; e.g.:

b   ... should be tested under XX conditions, because ...
... could perhaps be distinguished by ...
... would be desirable to try extending it into ...
... might therefore be interesting to ascertain whether ... is also true.
... if it could be proven that ...

If you have proposed multiple hypotheses regarding the cause of a newly discovered phenomenon, consider sketching out specific experiments for establishing whether one or another is in fact "correct" (or at least consistent with additional evidence), although a proposal of this sort could also be reserved for a subsequent "Conclusions" section; cf. Unit 13.

Subdividing the Discussion   You will almost certainly want to divide your Discussion into several subsections, if for no other reason than the fact that this chapter tends to be rather long. A few relevant structural considerations have already been put forward in Units 3 and 8. Unfortunately, there is no way we can offer a broadly applicable "recipe" for you to follow. The diversity in thesis content and scope is simply too great. Several organizational approaches might actually have merit. Often the most convenient framework from the perspective not only of the author but also the reader is one consistent with the structure of the Results section.

An unconventional outline?   The discretion you enjoy with respect to organization may extend even to the point of dispensing with the heading "Discussion" altogether. For example, with a highly theoretical project there might not be any justification for a section entitled "Results", which in turn implies there would be nothing obvious for you to "discuss"! Instead, you might want to direct attention toward examining theoretical point(s) of departure, or various model-building ventures, mathematical derivations, and/or attempts you launched at practical verification. A Results/Discussion formalism would thus be abandoned entirely. What we have treated as "standard structure" (cf. p. 54) often proves especially inhospitable to physics projects, even ones in which laboratory experiments are at the forefront. Typically the solution is an outline in which the equivalent of "results" and "discussion" coalesce into a single entity placed near the beginning. If you are tempted to consider a distinctly independent, personal approach to thesis structure be sure to discuss the idea with your advisor, after first taking into account any relevant university regulations.

A reminder — Keep in mind, incidentally, that "Results" by definition fall in the category of facts, whereas a "Discussion" has more in common with opinion—and in science it is always dangerous to mix the two. Your job as an author will most assuredly not become easier if you dispense with the conventional clear separation.

Significance — Of paramount concern in a Discussion is setting your findings in the proper perspective, which entails emphasizing—just as you presumably will in the abstract as well—the novelty and significance of your accomplishments. Returning for a moment to our peppermint-oil illustration (Ex 12–5), the researcher in question might want to underscore that now an effective new method—enantiomeric separation and analysis—is available for distinguishing natural materials from ordinary synthetic preparations, and that a scale-up, to preparative levels, of the analytical procedure just described could open the way to fragrance applications exploiting the more potent pure enantiomers (cf. also Unit 13, however).

Novelty — On the other hand, too much emphasis on the issue of novelty can become problematic. Suppose your contention that you are reporting something for the very first time is unexpectedly refuted at the eleventh hour by one of the members of your thesis committee—a potentially devastating blow, not only for you but also for your advisor! This ominous prospect is one of the reasons scientists generally tend to tread a bit lightly in their assertions:

Ex 12–9   … has never previously been accomplished in this particular way.
… is thus conclusively demonstrated, apparently for the first time.
… has to my knowledge not been successfully achieved in the past.
… is nowhere described in literature to which I have had access.

Questions of tense, etc. — Note that the reservations expressed in the last two examples almost of necessity introduce first-person language. Personal intrusion like this on the part of an author provides us with an opening for looking more closely at choices having to do with modes of expression in general. Broadly speaking, present-tense formulations should predominate in a Discussion section. After all: your main objective throughout your research has been to ascertain and better understand what *is* and what *is not*—to observe, to document, and to appraise, as reflected in

Ex 12–10a ... produces a sigmoid curve (Fig. X) ...
... is less pronounced than in a control group (Table Y) ...
... varies significantly with ...
... is therefore subject to ...
... reacts (forms, decomposes) ...

Past or past perfect constructions are more appropriate only when making direct reference to bygone events; e.g.,

b ... it did not prove possible ...
... had previously been avoided by ...
... it seemed necessary therefore to rule out ...
... Upon repeating the procedure under Y-conditions it became apparent that ...

Just as a reminder, the *past* perfect form ("it *had been* claimed") signals events or circumstances of a strictly historical nature, whereas something described in the *present* perfect ("it *has been* taken for granted") is understood to persist even now.

"The author speaking" With few exceptions we recommend you avoid expressing yourself in a thesis in the first person (apart, perhaps, from the Preface, or on an Acknowledgements page; cf. also Ex 12–9.) Even where personalized assessment of a result cannot be entirely avoided, we think it preferable to choose formulations that subtly imply your *hope* that something is true. In other words, instead of

Ex 12–11a I attribute this to ...
... in my opinion is ...

lean rather toward

b This in turn suggests ...
... can therefore (apparently, in no event) ...
... may thus be (assumed to be, regarded as, ignored) ...
... is for this reason likely to prove ...
... seemingly causes ...
... can most logically be attributed to ...
One may tentatively conclude ...

What matters in the long run is of course not what you currently *believe*, but rather what from your vantage point seem to be decisive factors. Present your arguments in such a way that others are encouraged to concur on their own, eliminating the need for you to intrude directly into the presentation, or to launch an overt appeal to your readership. There are in fact many ways of intimating in a gentle way the extent to which you consider results to be conclusive or far-reaching.

"We"   Finally, it goes almost without saying that you should under no circumstances in a thesis resort to first-person *plural* ("we") constructions. This would not only be utterly inappropriate; it could even plant disturbing suspicions in the reader's mind! (No such problem of course exists in a publication attributed formally to two or more authors.)

C 12–1   What might be said to epitomize the Discussion section of a thesis? In what sense are Results and Discussion sections expected to play different roles? (cf. also C 11–1)

C 12–2   Cite a few of the objectives you should set for yourself in the Discussion section.

C 12–3   How can you clearly differentiate between a description referring to your own results and one based on the work of others?

C 12–4   Under what circumstances in the Discussion might use of the simple present tense be justified, and when should you instead rely on (present or past) perfect constructions (see also Ex 11–4)? Illustrate your answer with examples.

13 Conclusions

- This unit deals with the potential value of including in your thesis a separate section for "Conclusions".
- Reading what follows will help you decide whether or not you should consider preparing a Conclusions section.

Q 13–1 Is there a section explicitly labeled "Conclusions" in most theses?

Q 13–2 How appropriate is it for such a section to incorporate speculation?

 2.2.8, 2.2.9

The "bottom line" Possible long-range consequences of results from a thesis research project are often explored in a separate chapter after the Discussion entitled "Conclusions". In a way, this could be interpreted as a "finishing touch": a place for you to comment on the overall significance of your findings. Put somewhat differently, a Conclusions section represents a sort of "bottom line" for your work, to borrow an expression common in the world of business to describe an overall profit/loss balance. There would be little point here, however, in simply reiterating (perhaps in different words) thoughts already presented in the Abstract. A Conclusions section should instead provide insights that are somewhat broader, but at the same time more limited as well. It does indeed possess some of the character of an overview, but it also casts glances toward the past and especially the future. Under no circumstances should you give in to the temptation in this section once again to put on display your entire project, methodological aspects included.

Editorial boards of some professional journals insist that every contribution they publish contain an explicit heading "Conclusions", but others argue just as strongly that "conclusions" should be regarded as a necessary (or at least logical) component to any formal Discussion. The same lack of consensus is apparent with respect to theses and dissertations: a section entitled "Conclusions" is no stranger to these documents, but it is by no means universal. Still, we must acknowledge that from a reader's standpoint such a chapter can be most welcome. A well-crafted collection of conclusions is a promising place to encounter the researcher at his or her creative best, and to savor a bit of intellectual stimulus while acquiring a clearer sense of a project's true value.

Consider the following example of a statement like that typically found in a thesis Abstract or Discussion (or both):

Ex 13–1a   … X has thus been prepared—in yields as high as 88%—in a unique medium: aqueous ethanol as opposed to the usual benzene …

Elaboration, generalization

This of course supplies a clear-cut "Yes!" to a fundamental question presumably posed at the very outset of the research regarding whether such a synthetic approach might be feasible. In a Conclusions section, however, the author might feel moved to elaborate and generalize a bit, with something like

b   … This new approach could prove to be of considerable commercial interest, since from the standpoint of workplace safety, as well as from an environmental perspective, water/ethanol as a reaction medium is far more attractive than benzene …

Speculation

One could even head off in a more speculative direction:

c   The yield in the process could probably be increased further by … , which would in turn reduce the reliance on ethanol …

Here we offer a few more samples of observations that would be very much at home in a Conclusions section:

Ex 13–2a   … For the first time, Y is now accessible from a remarkably simple precursor. As the Discussion has shown, this novel synthesis entails only three straightforward steps, and the resulting product, isolated in 78% yield, is almost entirely free of isomeric contamination …

b   … For this reason the new method should lend itself well to routine analytical applications, especially since every aspect is subject to straightforward automation …

c   … Elevated values for this parameter are evidently associated only with samples of type X. The implicit cause-and-effect relationship warrants prompt clarification, because one might be able to take advantage of it in a diagnostic context.

d   … so the addition of 0.2% ruthenium must have been responsible for the observed XXX. Taking into account the YYY factor, this would indicate that an even more dramatic effect is likely with rhodium.

"Conclusions" in this sense are always related in some way to implications and/or wider ramifications of a set of results, and with consequences that might perhaps be anticipated. Often they lead to specific recommendations for further research. Conclusions sometimes even assume the form of concrete predictions. Pursuing an earlier analogy, if everything in the body of a thesis prior to the Conclusions, including the Discussion, is equated with the telling of a story, then in the Conclusions the "moral" of that story is made manifest.

Under no circumstances should a Conclusions section be permitted to break new ground with respect either to the facts your research has uncovered or their basic interpretation, since these are matters reserved strictly for the Results and Discussion. You must also not allow the Conclusions to assume epic dimensions. Above all be sure whatever you write is not so bland as to be meaningless.

Given the nature of the challenge, consider building upon opening phrases like

Ex 13–3   It therefore stands to reason that …
This development would presumably lend itself to such applications as …
Only further study will resolve whether …

From another perspective, a Conclusions section represents a culmination of the Discussion, encouraging the reader to give explicit thought to the ultimate significance of your research efforts.

If conclusions do in fact find their way into an independent chapter in your dissertation, you might wish to consider restricting the introductory Abstract (Unit 9) more narrowly to a brief review of goals, methods, and results, stripping it of virtually all evaluative character.

C 13–1   Is it crucial that a thesis include a separate section entitled "Conclusions"?

C 13–2   Where would a Conclusions section best fit into a thesis?

C 13–3   What sorts of observations typify a good Conclusions section? Formulate a few concrete examples drawn from your own work.

C 13–4   Critique the following "conclusion" excerpted from the corresponding section of a masters thesis:

In summary, it may now be taken as established that, with various types of mammalian cells, alternative sera support growth much more effectively than fetal calf serum (FCS). This is not the case with all cell lines, however, so substitution of alternative sera must be considered independently for each cell variety studied. Generally speaking, however, alternative sera represent potentially interesting replacements for the typically more costly FCS.

14 Experimental Section

> ● Here we consider typical features of the "Experimental Section" of a thesis in science.
>
> ■ The discussion that follows should assist you in conferring a form suitable for a thesis on your various notebook descriptions and interim reports, as well as in integrating the nearly indispensable experimental dimension into your thesis.

Q 14–1 What level of detail is appropriate for the "Experimental Section" of a thesis?

Q 14–2 To what extent should your account of how you determined a specific value for some quantity include raw data (in conjunction, of course, with calibration and conversion factors, etc.)?

Q 14–3 What information would one expect to find in the Experimental Section regarding such things as sources of chemicals and other materials, apparatus, sampling procedures, and so forth?

Q 14–4 How do the Results and Experimental sections of a thesis differ?

Q 14–5 Is it common to find statements like "… the residual solution was not examined further …" or "… no investigation was carried out on the nature of byproducts …" in the Experimental section of a thesis?

 | 2.2.10 |

The "essence" of a thesis

Since the natural sciences are based almost by definition on experimentation, it is hardly surprising that most theses in scientific disciplines include sections devoted explicitly to experiments. With work that is largely theoretical in nature a comparable role could be fulfilled by a chapter entitled "Computational Methods", whereas the counterpart in a more descriptive setting, including medicine, might be dubbed "Experimental Subjects and Methodology"—or "Field Work" in the case of a geological, biological, or ecological study.

What experiments should be included? | Only experiments with a significant bearing on disclosures in the Results section need be included here. Details of other activities that may have occupied your attention can essentially be ignored. Each experiment—typically drawn from a single page (or a pair of facing pages) in a laboratory notebook (cf. Unit 1)—should lay claim to its own paragraph documenting the relevant procedures. A close relationship of course exists between the Experimental section and the two chapters "Results" and "Discussion", because you will be revealing here precisely how the various "answers" dealt with in these important narrative parts of the thesis were actually obtained.

Reproducibility | Describe each experiment in sufficient detail so that the reported outcome would be readily reproducible by anyone experienced in your general field of research. In effect, your job is to compile a "cookbook", although the writing style will differ somewhat from that of a typical laboratory manual. The sense of the presentation should always be "I did such-and-such; if you do the same thing, then you will obtain exactly the results I did, just as described here—and had you been here on the spot, you would have seen precisely what I have reported seeing".

Source materials | How rapidly you are able to commit this section effectively to paper will depend almost entirely—just as with the Results—on the quality of your source materials, in terms of their comprehensiveness and lucidity. The key sources are of course your research notebooks and various interim reports.

The experimental section places few demands on the author with respect to structure or planning, beyond simply establishing the sequence in which experiments will be covered. The arrangement should be based on logic, by the way, not chronology (cf. also the remarks concerning structure in Unit 12). A typical outline might take a form like the following:

Ex 14–1a
5   Experimental Section
5.1   Starting Materials
5.2   Procedures Derived from the Literature
5.3   Analytical Techniques
5.4   Reactions of X
5.4.1 Reaction of X with A
5.4.2 Reaction of X with B
5.4.3 Reaction of X with C
5.4.4 Reaction of X with D
5.5   Attempts to Optimize Z
...

or even the somewhat simpler

The latter scheme entails a bit less logical organization, sacrificed in the interest of eliminating one level of subdivision. Often (as above), a separate heading appears for each experiment. If you choose not to follow this course, however, it is imperative that the corresponding technical objective be spelled out explicitly in each paragraph's introductory sentence as an aid to the reader seeking a particular procedure (recall the broader recommendation in Unit 4 that *every* thesis paragraph should begin with a topic sentence); e.g.,

Ex 14–2a   **ACE Test:** Dansyltriglycine was …

b   In order to *determine the activity level of anti-factor-Xa*, a solution of …

(The definitive phrase in the preceding example has been set in italics solely for the benefit of readers of this book; one seldom sees text emphasized through italics, underlining, or boldface type in an actual thesis.)

Serial studies   If the need arises to discuss a set of several equivalent experiments, provide complete details for one representative example (to serve as a prototype), and then simply make note of such procedural changes as may be applicable to other cases; e.g.,

Ex 14–3   … Compounds of type **3** were prepared in most instances by a method strictly analogous to that reported in Sec. 5.4 for **2a** (yield of **3a**, 65%; **3b**, 70%; **3c**, 68%). In the preparation of **3b**, however, it was necessary to increase the standard reaction time (to 5 h) …

Experimental details   You should definitely make prominent mention of special technical considerations you feel may have affected the outcome of an experiment, or extraordinary precautions you took, including ones so minor they would ordinarily be omitted from a paper for publication in a professional journal. In other words, it is perfectly appropriate in the context of a thesis to insert an elaboration like

Ex 14–4   … was applied with a brush …

rather than simply saying "was applied".

Procedures already published elsewhere need not be spelled out in detail, even ones that figure prominently in your work. Assuming you introduced no significant innovations, a straightforward reference keyed to the Bibliography (Unit 15) will suffice:

Ex 14–5 … (see [17]) …
… as described by Jones and Davis (1997).

Raw data Especially in the case of an undergraduate or masters thesis we recommend you incorporate *all* raw data you collected in the course of your work; e.g.,

Ex 14–6 … NaOH consumption: 3.24 mL ($c = 0.1$ mol/L, correction factor 1.037) …
… yield of dry, crystalline product: 277.5 mg (87.4%) …

As illustrated in Ex 14–6, derived values such as yields would be appended directly to the information actually collected. This will, for example, permit a meticulous reader easily to confirm the validity of your calculations. In complicated cases you should also supply formulas you used in your transformation of raw data:

Ex 14–7 … Starting with the results of potentiometric titration (cf. [7]), the concentration c_t (in mmol/L) of HAMS at time t (in s) is obtained using

$$c_t = \frac{10 \cdot (f_a \cdot a - f_b \cdot b)}{n} - c_0(\text{OH}^-) - X$$

where

a initial excess of acid (in mL)
b consumption of base (in mL)
f_a, f_b concentrations of acid and base, respectively (in mol/L)
n sample size (in mL)
X initial concentration of HAMS (10 mmol/L in each case)
…

Sometimes you may even wish to walk the reader through the steps involved:

Ex 14–8 … For example, from Eq. (5) it follows that with $X = 0.02$ mol/L
the corresponding value of Y is 3.18 mL, which with Eq. (9) in turn
leads to $Z = 23.58$ mL/(s mol).

Figures, Tables It is rare in an Experimental section to find figures or tables, which ordinarily are more at home as part of the Discussion. Only if the Experimental chapter is situated near the *beginning* of a thesis is there incentive for embellishing it somewhat with tabular or illustrative material (such as sketches of apparatus, or flow diagrams, for example).

Tense and voice With respect to choice of tense, there really is only one sensible option for most of the Experimental section: the *past* tense, perhaps together with a few present-perfect phrases. One is, after all, mainly discussing activities that have long since been completed. On the other hand, certain types of assertion clearly refer to the present as well, as in "The compound's infrared spectrum shows …".

Passive constructions

You will probably find that most of the experimental descriptions you prepared for interim reports can be recycled directly in your thesis. In Unit 3, in the context of preparing interim reports, we advised against making use of first-person expressions, ones requiring "I" or "my". We feel equally strongly about the unsuitability of such forms with a thesis, especially in the Experimental section. Resist any temptation to say "I measured …" as a substitute for the staid, detached "… was measured". Apart from avoiding the monotony that can accompany repeated use of "I" (or "we" in the case of a multi-author document), passive construction offers the advantage that *things* play the role of subject in your sentences, thereby rising to the status they deserve. There is also no call here for trying to wax poetic. Readers expect to put up with a multitude of bland, impersonal, passive constructions in this section, although you should try not to build every sentence according to the tiresome "was done" model. Alternative formulations supply welcome variety; e.g.,

Ex 14–9   It proved advantageous to …
This was accomplished by …
In this way it became possible to …

The prose style of the traditional experimental write-up is notorious for its lack of glamour or embellishment. Much of the phrasing tends to be quite "routine" and essentially fixed. Such text is also likely to be littered with pieces of numerical data, arcane abbreviations, and parenthetical expressions. In the latter category are things like sources of starting materials, standards of purity, and indications of various experimental conditions (e.g., temperatures), all of which add essential bits of information, but are best marginalized such that they interfere as little as possible with the overall text flow. Make a point of closely examining the experimental details in recent articles from several professional journals to acquire additional insight into how your own work should be adapted for wider dissemination.

C 14–1   Develop a few subheadings you think might be appropriate for the experimental section of your thesis.

C 14–2   What sorts of information is it most important to incorporate into this section? How should you decide whether a particular experiment warrants coverage?

C 14–3   In what sense is derived data different from raw data? Illustrate your answer if possible with examples from your own work. Do both types of data deserve equal prominence? Why might you wish to illustrate certain transformations of raw data into derived data?

C 14–4   What elements in the following experimental descriptions could be criticized as imprecise, ambiguous, inadequate, etc.? Do any of the statements appear to belong instead in a Discussion section? What changes would you suggest, and why?

a   **4.3 Desorption of trichloroethene from polystyrene**

Desorption measurements can be made in a number of ways. One method involves determining loss of sample weight as it occurs during the actual desorption process. This is accomplished by placing the sample of interest in a glass container and hanging it from a spring balance. Weight loss can then be followed directly, in real time. Analytical procedures of this sort are described by Crank and Parl,[13] and also by Stuart.[15]

b   **Preparation of XXX**: A solution of AAA (moisture sensitive) in 30 mL BBB is warmed to 90 °C. A yellow precipitate presently appears, and this is filtered and dried (yield: 80%)

15 Bibliography

● This unit deals with the important issue of how background literature is to be cited in a thesis, and how one compiles the sources into a formal reference list.

■ You will find that literature references can take several different forms, and alternative approaches exist for identifying source material and assembling the information for presentation. The suggestions we provide should be of considerable help as you set about creating a proper bibliography for your thesis.

Q 15–1 What are two different (relevant!) meanings of the verb "to cite"?

Q 15–2 How are references to the literature incorporated into running text?

Q 15–3 What is a citation number?

Q 15–4 What principles underlie the name–date system of referencing?

Q 15–5 Is there more than one type of bibliography?

Q 15–6 What elements are most important in a source identification?

 | 9.1 through 9.5 |

Sources

A typical thesis includes a great many references to "sources" of information—often well over a hundred. Such references direct the reader to sites in the published literature furnishing information pertinent to the subject under investigation. Literature references play an especially important role in the section usually called the Introduction (cf. Unit 10).

Abbreviated references ("citations")

Citation links to the published literature should be made as unobtrusive in a document as possible, in the interest of minimizing their potential for distracting the reader. They must nevertheless provide efficient, unambiguous access to every source upon which the author relied. Incidentally: the word "citation" is actually used in two different, closely related ways. On one hand, "citation" can refer to the *act* of calling attention to a pertinent literature source, but a detailed *description* of that particular source is also known as "a citation" (hence "citation numbers"; see below).

Citation numbers The most compact approach to literature citation involves assigning sequential numbers to one's background sources as each becomes relevant in the master document. A correspondingly numbered Bibliography at the end is then used to convey the precise significance of each number. Citation numbers can be introduced into running text in various ways. Sometimes they take the form of small superscripts, but many prefer setting them directly on the text line, enclosed in either parentheses or square brackets. Both approaches require careful consideration of optimal text placement for the requisite number; e.g.,

Ex 15–1 ... as has been demonstrated with AAA,[4] BBB,[5,6] and CCC.[7–10]
... in contrast to previous reports.[3–11]
... agrees to a first approximation with published values [14, 15].
... similar to findings already announced for XX [9] and YY [10].

Superscript citation numbers are almost never preceded by blank space. They normally are introduced *after* adjacent punctuation (periods, commas, etc.), although the "em-dash", used to signal a pause or break, constitutes one important exception. Equivalent placement rules apply to superscript footnote markers (see Unit 17).

Authors' names It has become a rather common practice (welcomed by many readers!) in conjunction with numbered citations to mention explicitly—at least occasionally—the author(s) of the source in question:

Ex 15–2 ... as Jones [6] earlier demonstrated ...
... was recently shown by Smith and Partridge [7] to correspond to ZZ.

Note that it is never considered appropriate to refer in this context only to one party in the case of a source with several authors.

Name–date system Some graduate programs—and a wide range of publishers—actually prefer that in-text citations *always* identify the corresponding authors by name, and in such a way that citation numbers become superfluous. Particularly when one's bibliography is extensive, pursuing this strategy frequently requires also indicating a year of publication in conjunction with the relevant name or names in order to achieve specificity. The examples that follow illustrate what has thus come to be known as the "name–date system" of referencing:

Ex 15–3 ... Davis (1997) and Aldrich (2002) claim that ...
... is said to be substantially larger (Martin 1985; p. 132), although ...
... was first postulated (Miller and Lang 1997a) in conjunction with ...

Unlike superscript citation numbers, parenthetical citation expressions of this sort—as well as numerical citations, if these appear on the text line in brackets or parentheses ("fences")—always *precede* punctuation marks like commas and periods. The fenced expressions are then separated from surrounding text (but not from punctuation) by blank spaces.

Bibliography No matter which citation system you select, within your bibliography each reference entry is to be assigned a line (or lines) of its own. Indeed, comprehensive documentation of a source frequently consumes more than a single line. Just as with tables and figure captions, it is customary to use narrower line-spacing *within* a reference than that employed for body text, although extra space (equivalent perhaps to half a text line) is often introduced between individual entries in the interest of legibility. As always, you must obviously pay strict attention to any applicable university guidelines.

When the numerical system is adopted, every bibliographic entry begins with the corresponding reference number, generally again set as a superscript and/or in parentheses. This is followed by the reference itself, separated from the number by two or three blank spaces; e.g.:

Ex 15–4 a ...
[5] Kleine-Natrop HE. Derm Mschr. 1974; 160: 882.
[6] Devitt H, Clark MA, Marks R. Anal Biochem. 1978; 84: 315–318.
[7] Gloor M. Immobilized pH gradients. Amsterdam: Elsevier; 1990.
...

b ...
[12] Steinfeld JH. Urban Air Pollution: State of the Science. Science. 1972; 243: 745.
[13] Polster J, Lachmann H. Spectrometric Titrations: Analysis of Chemical Equilibria. Weinheim: VCH; 1989.
[14] Bartels K. Abfallrecht. Köln: Deutscher Gemeindeverlag; 1987.
...

Both sets of examples above conform to what is known as the "Vancouver Convention", according to which special roles (elaborated somewhat below) are assigned to the punctuation marks period, comma, hyphen, colon, and semicolon.

Name–date system As previously noted, there are no citation numbers in the name–date reference system, bibliographic entries instead being organized exclusively according to a scheme that is largely alphabetical, based on authors' last names (which in a corresponding reference list always precede the relevant initials). Multiple references attributable to the same author are arranged chronologically, oldest first. If more than one such source was published in a particular year, letters (a, b, c, ...) are appended to the year for differentiation purposes.

With multi-authored works the situation is a bit more complicated. If only two authors are involved, all such references are introduced immediately following any entries for the first ("principal") author alone, in alphabetical order by second author. Works with more co-authors come next, simply arranged chronologically. All this is actually much less complicated than it may sound, as should be apparent from the following fragment of a name–date author list:

Ex 15–5   Schmidt J. 1985.
Schmidt W. 1979.
Schmitt HP. 1986.
Schmitt HP. 1988.
Schmitt HP, Hinz A. 1985.
Schmitt HP, Kunz P. 1983a.
Schmitt HP, Kunz P. 1983b.
Schmitt HP, Kunz P. 1986.
Schmitt HP, Kunz P, Hinz A. 1980.
Schmitt HP, Hinz A, Fischer B. 1986.
...

Year of publication   It is important in a bibliography constructed according to the name–date system that the year of publication be situated relatively near the beginning of each entry—typically immediately after the name(s) of the author(s)—because of the crucial organizational role this information sometimes plays. By contrast, the year is often the *last* piece of information supplied under the numerical system, or at least one of the last—as illustrated in Ex 15–4a and 15–4b.

Editors and others in charge of publications have devised (and then made mandatory!) countless sets of mutually incompatible rules and conventions for structuring bibliographic references. As you plan your thesis we advise you to examine closely the bibliographies in two or three analogous theses that have recently met with the approval of your academic department, and subsequently adopt one of them as a model.

We begin a closer consideration of the way bibliographic data are reported by summarizing some of the more important constituent elements, arranged and punctuated as they might be in a reference list.

– In referring to an article from a professional journal:
Author(s). Journal title. Year; Volume number: Page.

– In referring to a book:
Author(s). Book title. Location of publisher: Publisher; Year.
(In the event the reference is to a specific passage in the book, its whereabouts should also be noted: by chapter, section, or even page.)

It is important to reiterate, however, that one encounters numerous variants on the illustrated patterns: from standpoints of style, organization, and punctuation. Periods are increasingly being dispensed with as signs of abbreviation, since many editors regard the period as better suited to separating the various discrete elements within a reference one from another. Also, a strong trend is apparent toward setting the author's initials *after*, rather than before, his or her family name—again, without periods. This is especially advantageous in references based on the name–date system, since it emphasizes the feature chiefly responsible for the list's alphabetical structure. Thus, Isabella Stuart James is transformed into:

Ex 15–6   … James IS (or occasionally the more traditional James, I. S.)

Names (titles) of journals are almost always abbreviated, and on the basis of an official, internationally sanctioned system. The most common abbreviations for this purpose can easily be deduced by examining reference lists from a few papers in the recent literature. Examples include:

Ex 15–7   J   for Journal
Z   for Zeitschrift
Exp   for Experimental

Here again, many editors refrain from the use of periods. Major words are typically *not* abbreviated, but spelled out in full (e.g., "cancer"), whereas minor words like "the" or "of" are simply ignored. Abbreviation is never permissible in book titles, however.

It is becoming increasingly common to find references to journal articles in which the precise title of the contribution of interest is also provided, often together with numbers identifying both the first and *last* pages; e.g.:

Ex 15–8   Reiter MB. Can you teach me to do my own searching? Or tailoring online training to the needs of the end-user. J Chem Inf Comput Sci. 1985; 25: 419-422.

Notice, incidentally, the distinctive absence in this example of capitalization for all title words but the first.

Multiple authors   When several authors (editors) have contributed to the work to be cited, as many as six may well be acknowledged in the corresponding bibliographic reference. With more than six, however, only the first three normally appear, followed by the expression "et al." (for *et alii*, and others) as an indication that names have been omitted.

It is essential that you acknowledge in your Bibliography (or "Reference List") every published source upon which your work can be said to depend, and that every listed reference be cited explicitly somewhere in the text.

Readers have a right to assume you have actually examined all your sources—in their original form!—and that you have carefully read at least significant excerpts from them. Make absolutely sure your assertions regarding the content of a source are scrupulously accurate. If you in fact did not have direct access to some important original source, but wish nonetheless to acknowledge it, structure the corresponding reference in terms of whatever document you really did consult (an abstract, or a paraphrase by another author, for example); e.g.:

Ex 15–9   ... (cited in [7]) ...
... (cited in [13]) ...
... (acc. to Chem. Abstr. 1991: 43215n) ...

Comments   Occasionally there is reason not only to cite a source, but also to comment on it; e.g.,

Ex 15–10   [17] Meyers C. J Fish Parasitol. 1981; 35: 285–291.— Most of the reported discoveries were made during the second Antarctic voyage of the "Meteor".
[18] Kunert F. Lipids. 1989; 72: 512–517.— Approximations to values presented here were previously disclosed by Eggert and Knut.[5]

If you adopt this practice, however, the resulting list would be better referred to as "References and Notes" or something analogous, not simply as a bibliography.

C 15–1   The following is an excerpt from a bibliography with numbered entries:

1.) Dr. Meier, Fresenius Zeitschrift für Analytische Chemie Volume 245.
2.) Hans-Peter Müller and R. Hausbold, Proc. of the Royal Chem. Soc. London, 134 (1980).
3.) V. R. Meyer Practical High-Performance Liquid Chromatography. John Wiley.
4.) Jander-Blasius, Laboratory course in inorganic chemistry.
5.) Prof. P. Ernestino, lecture notes.
6.) M. Princeling, M.S. Thesis, Georgia Institute of Technology.
7.) Pákányi et al. Cryst. Struct. Commun. 7 (1978) 435.

What features strike you as deficient or objectionable? How might the list be made more "modern"?

C 15–2   Below is an excerpt from a bibliography arranged according to the name–date system:

Y Bard, 1974. Nonlinear Parameter Estimation. New York: Academic Press. p. 145.
Milow, M. (1984), Talanta 1083.
M. Milow, Inorg. Chim. Acta 26, 947 (1984).

Milow, M. (1980), Talanta 1037–44.
Nagano K., Metzler, 1967. J. Amer. Chem. Soc. 89, 2891.
Polster J., 1975, Z. Physikalische Chemie Neue Folge 97, 113.
J.E. Ricci, 1952. Hydrogen Ion Concentrations. Princeton, NJ: Princeton
   University Press.

Identify any problems you detect.

C 15–3   Arrange the following (hypothetical) citations according to the rules of the
         name–date system.

a   Mayer PE, Zurman Z. 1983. J Amer Chem Soc. 83:1715–22.
b   Mayer PE, Schwarz AB. 1986. J Biol Chem. 258–61.
c   Mayer PE, Müller S, Schwarz AB. 1991. Biochem Biophys Res Commun.
    95: 1752.
d   Mayer PE, Schwarz AB, Müller S. 1986. Angew Chem Int Ed Engl. 25: 429.
e   Mayer PE, Müller S, Schwarz AB. 1987. Talanta 33:514.
f   Mayer PE. 1968. Photoluminescence of solutions. Amsterdam: Elsevier.
g   Mayer PE, Schwarz AB. 1987. J Amer Chem Soc. 86: 86–8.
h   Schwarz AB, Mayer PE. 1986. Z Naturforsch. 41a: 1350.
i   Schwarz AB. 1990. Biochem J. 82:112.
j   Mayer PE. 1968. J Chem Educ. 45: 312.

C 15–4   What document types are you familiar with as potential sources that (in
         the interest of brevity) have not been explicitly addressed in this unit?

C 15–5   Has sufficient information been provided here to enable you properly to
         cite all types of books?

C 15–6   Would you consider it appropriate to cite without comment an entire book
         if it contains only one relevant passage? If not, what would be a better
         course of action?

C 15–7   Suggest a reasonable way of handling the following situation:

         Smith H and Johnson MR have published an article you wish to cite. It appears
         as pp. 107–117 in Volume 20 of a book series edited by B Shulz with the series
         title "Insects of the Amazon Rain Forest". The volume in question was released
         by World Press in Tallahassee, Florida, in 1990.

C 15–8   The following text excerpts are derived from theses. Suggest possible im-
         provements with respect to the placement of literature citations.

a   … became accessible through a "half-sandwich" structure[45]; apart from HX,
    numerous other electrophiles containing sulfur[46], selenium [47] and
    tellurium[48–50] as the key atom—the same applies as well to carbenes[51,52] and
    nitrenes—[53] are subject to addition, as are certain metal compounds with
    Lewis-acid character, including CuCl, among others[54].

b   … Müller[12], Kandroro, [13–16] Finnigan[17] and also Mertz [18] et al. have
    established that the process in fact consists of an acid-catalyzed isomerization
    of **20** to **21**; complexes with linear[19] building blocks X—especially spectacu-
    lar: X = H–C≡C–H as ligand—[20] have also been prepared quite recently;
    [21–22] experiments with Y–C≡C–H (Y = Me)[23] were unsuccessful.[24]

16 Appendices, Miscellaneous Other Sections

- ● Here we discuss other miscellaneous sections sometimes found in theses.
- ■ The goal is to help you decide what (if anything) might usefully be relegated to an appendix, and to focus generally on concluding your efforts.

Q 16–1 What sorts of information are best dealt with through appendices?

Q 16–2 Apart from appendices, what other miscellaneous sections might prove useful (or mandatory!)?

 | 2.2.11

Appendices

Following the Bibliography (cf. Unit 15) it is sometimes appropriate to introduce a certain amount of additional material, especially in the form of one or more appendices ("Appendix A", "Appendix B", …). In the event that a proposed appendix will for some reason contain literature citations, however, it should actually be situated *ahead* of the Bibliography.

Accompanying material

Appendices can be the perfect place to accommodate extensive sets of comparable experimental results—but also spectra, flow diagrams, mathematical derivations, computer programs, photographs, and in an electronic thesis perhaps even videoclips, to cite but a few examples. All such elements have in common that their presence elsewhere in the document could significantly disrupt the flow of text. It is important to emphasize, however, that an appendix should *never* be made a repository for important revelations not otherwise unveiled in the main body of the thesis, in part because appendices are in fact rarely subjected to examination, so their content tends to be overlooked. The point can perhaps be underscored by noting that when the work culminating in a thesis is made the subject of journal articles, material from appendices typically ends up being excluded.

Multiple appendices *Multiple* appendices are worth considering when a need arises to deal with more than one type of information, or if a compartmentalization on the basis of topic seems desirable.

Remarks Some theses are distinguished by the fact that near the end one finds another special section, labeled "Remarks" and dedicated to supplemental discussion, clarification, observations, and the like—again, things that would represent a serious distraction or an unjustifiable diversion if grafted into the body of the thesis. On the other hand, what an author proposes to communicate in this way may be sufficiently limited in scope (frequently the case, especially in theses from the natural sciences and engineering) that it would better be transformed into a few appropriately placed footnotes (see Unit 17). You may also recall our brief mention in Unit 15 of the possibility of embedding a few notes in a (suitably titled) section whose primary purpose is actually to provide formal literature references.

Vita Another piece of "back matter" that could well be called for in your thesis is a "personal vita" (curriculum vitae). This is nothing more than a concise record of background information you would likely be asked also to supply in the context of a job application: essentially the high points (so far!) in your overall educational journey, as well as relevant professional experiences you may have had. A vita for this purpose should be confined to a single page, and might be of the narrative type or, alternatively, assume an outline form like the following:

Ex 16–1

Name	David Grove Smith
Place and date of birth	Abilene, Kansas, 14 July 1979
Primary schools	Washington Elementary, Abilene, KS, 1984–87; 14th Street Elementary, Topeka, KS, 1987–1991; Roosevelt Junior High, Topeka, KS, 1991–1994
Secondary school	John Sterling High School, Tulsa, OK, 1994–1997
Higher education	Texas Christian University, 1997–2001; Phi Beta Kappa, 2000, B.S. in Geology, *magna cum laude*, 19 June 2001 Dollinger Fellow, Department of Geology, Oregon State University, since September, 2001; Sigma Xi, 2002
Other studies	Foreign study participant, University of Lille, France, June–August, 1999 (fluent in French)
Employment history	Salesman, Wal-Mart, June–August 1997; Laboratory Assistant, TCU Geology Dept., September 1998–May 2001

Many institutions also require certain types of *lists* in submitted theses—to be inserted immediately following the Table of Contents: a list of all the tables present (cf. Unit 20), for example, and another of all figures (graphs, illustrations, drawings, etc.; cf. Unit 21), perhaps accompanied by a list of the symbols and/or abbreviations you have employed (cf. especially Unit 18). Similarly, don't be surprised if there is also a requirement for a separate page to serve as a copyright notice, and possibly even documentary evidence that you have obtained (from the appropriate sources) proper permission to reproduce material copyrighted by others.

Signature page   Finally, it is quite common for schools to expect theses to display a formal "signature page", a place where one's adviser and the other members of the thesis committee can testify in an official way to their formal approval of the work. Like the Title Page (Unit 6), this Signature Page will probably have its layout and structure specified in great detail by university authorities. In most cases a Signature Page would be placed at the very beginning of the document.

C 16–1   A thesis dealing with anatomical aspects of a particular zoology project entails, among other things, careful quantitative interpretation of a host of photomicrographs derived from tissue thin sections. Important parameters have been derived from scores of painstakingly established numerical values, with considerable help from a unique computer program the author developed. All the relevant material is to be compiled in appendices, with only selected examples and summary information appearing in the body of the document. Suggest appropriate structures for one or more appendices that could serve this purpose.

C 16–2   Identify techniques commonly employed to obviate the need for a separate thesis section devoted to "Remarks".

C 16–3   Try assembling the text for a personal vita suitable for inclusion in your own thesis: *without* first reexamining Ex 16–1. When you finish, compare the results with our example. Do you see things you in fact wish now to change?

Part III
Special Elements

17 Footnotes

- The present unit explores ways in which footnotes can prove to be much more than simply a bothersome formality.

- After examining this material you will find yourself well placed either to exploit the concept of footnotes to your advantage, or avoid them entirely.

Q 17–1 What sorts of information are more suited to inclusion in footnotes than in running text?

Q 17–2 How does one ensure that footnotes are easily distinguished at a glance from the body of a document?

Q 17–3 Under what circumstances should you indicate the presence of footnotes with symbols rather than numbers?

Q 17–4 With what *disadvantages* are footnotes associated?

Q 17–5 What are strategies sometimes adopted to bypass the need for footnotes?

Q 17–6 Should literature citations take the form of footnotes?

Q 17–7 Notes associated with tables are also described as footnotes. What keeps these from being mistaken for running text?

 | 5.5 |

Annotation via footnotes

Footnotes represent a special form of annotation usually restricted to lengthy documents. In most cases their chief function is to present commentary the document's author regards as secondary relative to the principle message(s) to be conveyed. Brief asides in this category might simply be set in parentheses, or isolated with the aid of ("long") dashes, but footnotes have the advantage of being rather unobtrusive, so they pose less threat to the serious reader's concentration. The term "footnote" is derived from the fact that such added morsels always appear at the "foot" of the page: out of the way, but nevertheless conveniently nearby.

"Notes" as a separate section — The fact that they are set in close proximity to the text passages with which they are associated is one of the most important characteristics of footnotes. A less convenient alternative (especially from the standpoint of the reader) entails deferring notes to the end of the current chapter, or even to a separate section of the document, created for that express purpose and with a heading like "Supplementary Remarks", or simply "Notes". The comments are handled in this case as a collection of discrete, sequentially numbered paragraphs (generally referred to as "endnotes"). Yet another possibility is to merge any such remarks into what would ordinarily constitute the bibliography—which then of course requires a broader title: "References and Notes", perhaps (cf. Unit 15).

Footnote markers — For every footnote you choose to incorporate, some sort of marker must obviously be affixed to the most relevant text passage on the page in question, which can then function as a link. These markers often take the form of tiny symbols printed as superscripts, such as an asterisk (*) or a plus sign ($^+$).

Replication of a particular symbol — Replication is a permissible means of taking advantage of a particular symbol as many as three times on a single page (e.g.: *, **, ***) before one is forced to resort to a suitable alternative. "Special" symbols frequently employed as footnote markers include, in addition to the asterisk and the plus sign:

Ex 17–1 — †, ‡, §, #, ¶

In many cases *numbers* are instead used to call up the footnotes, but this option presupposes that numbers have not already been co-opted for directing attention to pertinent references to the published literature (Unit 15). When numbers are in fact utilized as footnote markers, some authors elect to revert to "1" at the start of each new page, while others number notes consecutively on a chapter-wide basis (or even throughout the document as a whole).

Appending a "close parentheses" symbol — Especially in scientific text it has become commonplace to append a "close parenthesis" symbol to in-text footnote markers to eliminate any chance of confusion with superscripts that serve other purposes (e.g., exponents). The following example illustrates several important structural conventions traditionally associated with the use of footnotes:

Ex 17–2 — Text text text$^{*)}$ text text text text text text text text text text text text text text text text text text$^{**)}$ text text text text text text text text text text text text

> Space ratio at least 1.5:1 (see below)

---

* Footnote footnote footnote footnote footnote footnote footnote footnote footnote footnote.

** Footnote footnote footnote footnote footnote footnote footnote footnote footnote footnote footnote footnote footnote footnote footnote.

The footnotes themselves are separated by one or two spaces from markers that precede them. Usually the notes are aligned consistent with body-text alignment (as in Ex 17–2), although sometimes it is instead indented in a uniform way:

Ex 17–3   * Footnote footnote.

** Footnote footnote footnote footnote footnote footnote footnote footnote footnote footnote footnote footnote footnote footnote footnote.

Footnotes and their associated markers should always be set in type smaller than that used for body text (e.g., 10-point rather than 12-point). Some prefer to employ for the purpose a different type font altogether. Fortunately, most of the task of footnote formatting can be delegated to a full-featured word processor (like Microsoft WORD), although an author always retains the power to specify a custom format of his or her own design.

It could perhaps be argued that details like the foregoing represent only arbitrary matters of "style", but (consciously or unconsciously) a great many readers take for granted the traditional practices, which lend an air of professionalism in the context of formal documents—including theses! The key to meeting inherent high expectations is taking full advantage of technical computer-based facilities already at your disposal.

Table footnotes   Tables constitute a special situation in which it often becomes virtually impossible to avoid introducing a few footnotes (see Unit 20). It is important in this particular case to steer clear of numerical markers, since these pose an especially serious risk of confusion with tabulated data. Superscript (lower-case) letters are in fact usually preferred, again followed by "close parenthesis" signs. As usual, the selected symbol is repeated—albeit without the accompanying ")"—in front of the footnote itself, which is placed immediately below the corresponding table (cf. Ex 20–4).

It is never appropriate to include footnote markers (or literature citations, for that matter!) in headings. Any temptation to do so can be alleviated by formulating the lead sentence following that particular heading in such a way that it quite naturally accommodates the desired link; e.g. rather than:

Ex 17–4a   **Header**[*]

Text text text text text text text text text text text text text text text text text text text text text text text text text text …

---

[*] This phenomenon was first described in a paper by Jones et al. [16].

build in a receptive opening sentence (or phrase) like:

b Header

XXX was first described in a paper by Jones et al. [16].

Text text text text text text text text text text text text text text text text text text text text text text text text ...

The in-text superscript link to a footnote (footnote marker)—set, as indicated, in type smaller than that employed for the text itself—should be placed as close as possible to the specific text element (word, phrase) to which that footnote most directly applies. Such marks belong *after* any adjacent punctuation, the sole exception being the long dash ("em dash") indicative of a pause; e.g.:

Ex 17–5   ... was encountered for the first time.[1] Later, ...
... according to some interpretations,[2] but ...
... should then be optimized;[*] in this way ...
... seems quite straightforward[+]—but in fact ...

The superscript numbers (reference numbers, citation numbers) required under the numerical reference system for calling attention to pertinent literature sources (cf. Unit 15) are of course links of an analogous kind, but with a very special purpose. These superscripts are also subject to the placement rules outlined above.

Proper page placement of footnotes

Footnote text should be separated from the last line of body text on a page by a horizontal line at least 20 mm long and extending to the right from the left margin. Uniform blank space is to be provided both above and below this dividing line, with that above at least 50% wider than the space below (see Ex 17–2). Ideally, the last line of every footnote block should fall exactly where, in the absence of footnotes, ordinary text would be expected to cease on that page. Adherence to rigid guidelines of this sort in times past—when dissertations were prepared manually using a typewriter—was a difficult and truly painstaking proposition, but now with the advent of word-processing software essentially all the desired ends can be achieved automatically, and almost always flawlessly.

Sometimes a footnote marker proves necessary so near the bottom of a page that insufficient space remains to print the footnote itself. In this case one still incorporates as many lines as possible of the note in question, followed either by an indication of where the remainder should be sought, e.g.,

Ex 17–6a   ... Footnote footnote footnote footnote footnote footnote footnote (contd. on page x)

or simply by an arrow implying that the reader should turn the page. Alternatively, the entire content of the offending note might be displaced to the subsequent page:

b   * Footnote see page x.

It is sometimes suggested that "note-continuation text" be distinguished from ordinary note text by being set under a dividing line that extends across the entire page.

With numbered footnotes it is important to ensure that, if marker placement shifts in the course of editing, no sequence errors are introduced. This is rarely a problem with footnotes managed automatically by a word processor, however, since the software is equipped to keep perfect track of note placement, and adjusts all numbers as required. In most cases word processors also carefully monitor page breaks, providing as necessary for proper note continuation (in the proper way) on a subsequent page.

Word processors permit the author to choose freely between true footnotes and the alternative of "endnotes" (for incorporation at the end either of a chapter or the entire document), and preferences in this regard can be changed at any time in the course of composition. The same capability clearly makes possible the analogous treating of numbered literature references as endnotes, which in turn permits essentially automatic formulation of one's bibliography.

C 17–1   Describe common alternatives to the use of footnotes.

C 17–2   What sorts of symbols are preferred as footnote and table-footnote markers? Why?

C 17–3   Identify problems that might arise with a manuscript containing both numbered footnotes and numbered literature citations.

C 17–4   How are footnote markers best avoided in headers?

C 17–5   Correct all errors you detect regarding placement of footnote numbers in the following simulated thesis excerpt:

Text text text text.[1] Text text text text text text[2], text text text text[3,4], text text[5]; text text text text text text[6–9], text text—text text text—[10] text text text text text text[11].

_____

[1] Footnote footnote footnote footnote footnote footnote footnote ...

18 Numbers, Quantities, Units, and Functions

- In this section we discuss the use of numbers, quantities, the units in which the latter are measured, and mathematical functions, together with related typographic issues.

- Studying this material will prepare you to incorporate measurements along with other quantitative information properly into your thesis.

Q 18–1 What is meant by the concept of a "quantity"?

Q 18–2 What significance do you attach to the term "SI"?

Q 18–3 How would you describe a "unit of measurement" in science, and what is the role of a "unit symbol"?

Q 18–4 How many so-called base units are there?

Q 18–5 Assuming a particular symbol requires both a subscript and a superscript, in what order should these appear on the page?

Q 18–6 Which of the following reflects proper notation: 22 °C, 22° C, or 22°C?

  6.1 through 6.5

Numerical data Numerical ("quantitative") information plays a key role in much scientific text. It is of course essential that the author of a thesis in science carefully check every piece of reported data to guarantee the absence of errors, including so-called typographical mistakes. This requires devoting particularly close attention to tables and figures, both in the course of careful proofreading and then again during a final scrutiny of the finished copy, including monitoring things like decimal points and applicable units of measurement. Be sure not to limit your inspection to verifying that data have been properly transferred from source to document: check also to make sure that all reported numbers in fact seem plausible! It is certainly preferable to carry out one calculation too many than one too few.

Precision   In reporting numbers, be conscious always of the appropriate level of precision (i.e., number of "significant figures"), which is a direct function of the way data have been acquired. Whenever possible, measurements should also be qualified by "± factors" as a way of acknowledging and adding specificity to unavoidable uncertainty. It is often necessary to apply conversion factors to data, and in doing so it is important that you consider the way potential errors are propagated. In all cases you should be explicit in identifying what meaning a reader should attach to the specification of a range. Reporting the total number of all measurements that have contributed to a particular numerical value is also essential, as is indication of the scope of random sampling that may have been performed.

Statistics   It is almost always a good idea to subject experimental data to regression or correlation analysis. (Further information can be obtained from any standard statistics text). Resulting significance or reliability levels are often most conveniently reported in footnotes to the tables typically employed for data presentation (cf. Unit 20). In the case of data displayed graphically, uncertainty can be communicated especially effectively with the aid of special symbols devised for that purpose (cf. Ex 21–17b). Most thesis writers today rely on PC ("personal computer") software to perform their statistical analyses, and to organize the outcome in tabular and/or graphic form. A word of caution is in order here, however: be sure you take advantage for this purpose of true *statistical* software (e.g., SPSS, MINITAB), not the notoriously less reliable functions offered by spreadsheet programs (e.g., Microsoft EXCEL).

Successful information transfer in the context of scientific documents has as a prerequisite extreme care in the use and treatment of symbols (e.g., "quantity symbols"). In sharp contrast to the drastic limitations faced by authors back in the era of typewritten theses, now virtually everyone is able easily to access a vast array of type fonts, with provision not only for standard "roman" characters, but italic and boldface styles as well. This is complemented by unprecedented flexibility in terms of type size and spacing. With a treasure trove like this at your disposal it is assumed you will demonstrate intimate familiarity with the fundamental typesetting rules specifically applicable not only to standard text but also mathematical "text" (i.e., formulas and equations). Especially important with the latter is the set of guidelines identifying circumstances under which you should use either upright ("roman") or sloped ("italic") type for expressing both numbers and symbols in a variety of contexts, as well as standards, depending on the situation, for proper spacing of letters, numbers, and other symbols. Considerations such as these contribute far more than you might imagine to the professional appearance of a document. Table 18–1 provides a convenient summary of the most important rules in the typesetting of mathematical and physical expressions and formulas.

Table 18–1. Basic typesetting rules applicable to expressions of the mathematical/physical kind.

Authorized stylistic treatment	Illustrations
Italic ("slanted") type	
Mathematical variables	$a, b, c, x, z, A, B, \alpha, \beta, \gamma$
Symbols for physical quantities	m, t, T, r
Symbols for general functions	$f(x) = u(x)/v(x), z = \varphi(x,y)$
Symbols for natural constants	R (ideal gas constant), N_A (Avogadro's number)
Roman ("upright") type	
Numbers	1, 2, 3, 2005, π, e
"Fences"	( ) [ ] { }
Operators	d, D, Δ, ∇, ∂, %, ‰, ppm, ppt
	$df(x)/dx$, $\partial g(x,y)/\partial x$, 2 %, 0.1 ppb
Operational symbols	$+, -, :, \times, =, <, >, \in, \cup, \cap, \Rightarrow$, AND, OR
Symbols for special functions	exp, log, ln, lg, sin, cos, tan, Re, Im
	$\cos x$, $\exp(-x^2)$, $\mathrm{Re}(z) = a + ib$
Unit symbols	m, kg, s, A, K, mol, cd, °C, W, V, Pa, ha
Unit prefixes	G, M, k, m, μ, n, p; nm, GHz, mbar
Summation, product, and integral signs	$\Sigma, \Pi, \int$
Extra space	
Within numbers	4150 17 315 1 247.014 33 3 1/2
Before and after operational symbols	$3 + 4 = 7, f(x) = x^2 - 2x$, 18 mm $\times$ 24 mm
Between numerical values and units	3 m 13 °C 180.15 K 12 mmol/L
Between terms in products of units	70 mg mm^{-1} L^{-1} 0.4 mg/(kg a)
With proportional symbols	12.4 % 0.1 ‰ 20 ppm

"Roman" vs. "italic" type

One fundamental principle to stress at the outset is that numbers (including individual digits, or numerals) in a mathematical environment are always to be displayed in ordinary roman type. This generalization covers even the unusual numbers one expresses in special ways, such as π, e, and i. Also to be set invariably in ordinary type are symbols representing the units with which quantities are measured ("units of measurement"; e.g., nm, kg, mol). Roman type is also the only correct medium for arithmetic signs, symbols derived from set theory or formal logic, operators (including the differential operator, cf. the first symbol in the series presented as Ex 18–1), and certain so-called special functions, like those in Ex 18–3:

Ex 18–1   d ∂ δ D Δ ∇ ppm ppb ppt % ‰

Ex 18–2   + – : $\times$ > < $\oplus$ $\otimes$ $\in$ $\varnothing$ $\cup$ $\cap$ $\exists$ $|$ $\Rightarrow$ $\Leftrightarrow$

Ex 18–3	exp log ln sin cos tan sinh cosh Re (real portion) and Im (imaginary portion): $z = a + ib$ $\mathrm{Re}(z)$ $\mathrm{Im}(z)$
Variables, quantities, and the SI	Mathematical *variables,* together with the accepted symbols for physical quantities (including SI—Système International d'Unités—base quantities, cf. Table 18–2), explicitly take the form of *italic* letters, drawn from either the Latin or the Greek alphabet.
Ex 18–4a	Mathematical variables: $a, b, c, x, y, z, \alpha, \beta, \gamma$
b	Physical quantities: m (mass), l (length), t (time), T (thermodynamic temperature)

Table 18–2. SI base quantities, with their approved symbols and units.

SI Base quantity	Quantity symbol	Corresponding base unit	Unit symbol
length	l	meter	m
mass	m	kilogram	kg
time	t	second	s
electric current	I	ampere	A
thermodynamic temperature	T	kelvin	K
amount of substance	n	mole	mol
luminous intensity	I, I_v	candela	cd

	Only in a few rare cases have *pairs* of letters been approved as symbols for physical quantities; e.g.,
Ex 18–5	*Re* (for the Reynolds number)
	Symbols for natural constants should also be italicized (after all, these represent physical quantities as well!):
Ex 18–6	N_A (Avogadro's number) h (Planck's constant) R (the ideal gas constant)
General functions	Moreover, letters used to designate "general functions", in contrast to the "special functions" touched on above, are set in italic type; e.g.
Ex 18–7	$f(x) = u(x)/v(x)$ $z = \psi(x, y)$
Vectors	Symbols for vectors and matrices, on the other hand, are to be displayed in *boldface* italic type; e.g.:
Ex 18–8	$\boldsymbol{a} = a_1\boldsymbol{e}_1 + a_2\boldsymbol{e}_2$ $\boldsymbol{A} = \begin{pmatrix} 1 & 0 \\ 0 & 2 \end{pmatrix}$ $\boldsymbol{F}$ (force) $\boldsymbol{E}$ (electric field strength)

It once was customary to indicate a vector by an appropriate quantity symbol topped with a small arrow (e.g., $\vec{a}$, $\vec{F}$, $\vec{E}$), but this notation is no longer recommended.

Tensors Yet another rule with respect to type states that *tensors* should be represented by roman, boldface characters taken specifically from a "sans serif" (i.e., unembellished) font; e.g.,

Ex 18–9 **A B C**

Specificity can be conferred on a letter symbolizing a quantity or variable by attaching to it one or more qualifying subscripts and/or superscripts ("indices"), or in some cases crowning it with a "hat" in the form of a straight line or other distinctive mark. For example,

Ex 18–10a C_p (both in italics)

is used to represent the heat capacity (C) of something at constant pressure p, just as

b g_n (g italicized, n roman)

is the accepted representation for the acceleration due to earth's gravitational force g, where the "n" here signifies "normal".

A clear explanation should be supplied when any symbol makes its first appearance in a document. It is strongly advised in addition that all symbols present, and especially those representing quantities, be meticulously defined in a separate "List of Symbols", which also identifies applicable units of measurement. This list should be positioned near the beginning of the work. Ex 18–11 is illustrative of information drawn from such a list:

Ex 18–11 c molar concentration of adsorbent in the bulk liquid phase, mol m^{-3}
C_L liquid-phase specific heat, J kg^{-1} K^{-1}
D_p adsorbent particle diameter, m

Apart from type style, another important consideration when preparing mathematical copy is spacing. Sometimes the introduction of extra space would class as merely "desirable", in that it offers the potential for improving legibility, but in certain situations—cf. m s^{-1} (meter times second to the minus first) and ms^{-1} (millisecond to the minus first)—close attention to spacing becomes absolutely prerequisite to the reader's unambiguous interpretation of an expression.

Especially in tables, and quite generally for numbers consisting of more than four digits, accepted practice is now to group the respective digits into *spaced triads* extending out in both directions from the decimal point; e.g.,

Ex 18–12   17 315   1 215.01   0.000 000 1   3 435 123.010 45

On the other hand, strictly uniform spacing is called for with numbers composed of four or fewer digits:

Ex 18–13   2003   + 9999   – 3300

A special narrow space is required between a whole number and any accompanying fraction:

Ex 18–14 $3\,^1/_2$

The symbol for a unit of measurement must also be clearly separated from the corresponding numerical value, this time by a (standard) single space. For example:

Ex 18–15   3 m   280.15 K   17.5 kg   12 mol/L   0.05 mg/(kg a)

A space introduced for this particular purpose should be of the "non breaking" type—one that is physically incapable of inadvertently being transformed into a line break—as a way of ensuring that a quantity's two constituent elements always remain together.

The symbol combination "°C" is considered a single discrete entity, with the meaning "degree Celsius". It should therefore—in its entirety!—always be set one space removed from an associated numerical value; e.g.,

Ex 18–16   13 °C (*not* 13°C or 13° C)

With the arguably analogous notation for angular degrees, however, the degree symbol is instead printed flush with the last digit of the corresponding number, a rule applied also to the symbols for angular minutes and angular seconds; e.g.,

Ex 18–17   180°   12.3′   51° 12′ 35.5″

Note that the symbol for angular second is ′, which one should never try to approximate by either an apostrophe or a single quotation mark (' or ').

Line breaks   As indicated above, it is important to be sure that numerical values never become separated from their associated unit symbols. Word processors offer provisions not only for the requisite non-breaking spaces, but also for creating spaces with various widths, which can in turn be either of fixed dimension or flexible enough to play a role in the adjustment process that accompanies preparation of fully justified text (cf. "Treatment of the Right Margin" in Unit 5).

Tolerance limits   Information related to error limits or "tolerances" should be conveyed using a notation like

Ex 18–18a (142 ± 10) nm

not the commonly encountered (but highly improper)

b 142 ± 10 nm

The latter's "impropriety" stems from the fact that the unit symbol applies to both the alleged error bounds *and* the measured value itself.

Range   A *range* is most frequently signaled with assistance of a dash, as in 800–1000 bar (which actually should take the form of an "en dash", a character noticeably longer than a hyphen and comparable in length to a minus sign; cf. Unit 19). This practice suffers from introducing some risk of confusion with the true minus sign, however, so in recent years an alternative symbolic approach has been promoted:

Ex 18–19   800…1000 bar (without parentheses)

Again, a single (non-breaking) blank space should be inserted between a numerical value and the symbol used to indicate either "percent" or the less familiar "per mill",*) as well as in conjunction with the related expressions ppm ("parts per million"), ppb ("parts per billion"), and ppt ("parts per trillion"); e.g.,

Ex 18–20   12.4 %   0.1 ‰   20 ppm   0.05 ppb

It is worth noting, incidentally, that the "ppb" specification can be a source of ambiguity, since in most cultures "one billion" is interpreted as equivalent to 10^{12}, not 10^9 as in the United States.

SI units   The following symbols have been officially assigned to the seven "base units" recognized under the SI (cf. Table 18–2):

      m   kg   s   A   K   mol   cd

Prefixes   The range of each of the base units (and other units as well, including the "ton", symbolized by "t") can be extended or truncated by successive factors of 10^3 through use of certain authorized prefixes, each of which can also be expressed symbolically. Prefixes in this category include mega (M, 10^6), kilo (k, 10^3), milli (m, 10^{-3}), micro (μ, 10^{-6}), and nano (n, 10^{-9}). The appropriate prefix symbol is placed directly to the left of the impacted unit symbol; e.g.,

Ex 18–21   Mt (megaton)   kg (kilogram)   mmol (millimole)   nm (nanometer)

It is never permissible, by the way, to write such a prefix in isolation, for instance as a "convenient" substitute for the corresponding power of ten itself.

h, da, d, c   Use of the once common prefixes hecto (h, 10^2), deca (da, 10^1), deci (d, 10^{-1}), and centi (c, 10^{-2}) is now discouraged, although it seems highly unlikely that the unit "centimeter" (cm) will be abandoned any time soon.

---

* The percent sign is sometimes considered to be an exceptional symbol that is *not* separated from the associated number (and it is typically so treated in the text of this book), or separated from it only by a thin space.

By the way: in expressing a measurement (a "quantity") as the product of a numerical value and a unit you should always try to adjust the unit (through a suitable prefix) such that the resulting numerical value will fall between 0.1 and 1000. In other words, it is better to write

Ex 18–22   30 μL

rather than the equivalent 0.030 mL.

As implied earlier, a single (non breaking) blank space is used to separate a pair of unit symbols to be understood as representing a product; e.g.,

Ex 18–23   70 A s 10^{-2} g/(m^2 d)

In the second example above, note particularly that the product of units occurs in the denominator (i.e., following a slash), and it has been enclosed in parentheses as a way of eliminating ambiguity that would otherwise exist (as elaborated on p. 146).

Indices, incl. exponents For the sake of appearance, subscripts and superscripts should be set in type roughly 3/5 the size of that used for the corresponding host symbol. In other words, a 7 pt or 8 pt font would be an appropriate choice for indices associated with a symbol displayed in 12 pt type.

Subscripts and superscripts belong immediately adjacent to the symbols they are intended to modify. If the symbol for a variable requires both a subscript *and* a superscript, the former should if possible be set directly above the latter. Following this recommendation may require you to take advantage of special "mathematically-savvy" software (such as the equation editor supplied with Microsoft WORD; further information is provided in Unit 19). For example:

Ex 18–24a K_S^2 n_{max}^2

If you find yourself unable to comply with this standard, alternatives such as the following are available:

b K_S^2, $(K_S)^2$ or n_{max}^2, $(n_{max})^2$

In the case of chemical formulas for charged particles (ions), one of the latter typographic solutions is actually deemed preferable; e.g.,

c NH_4^+, SO_4^{2-} or $(NH_4)^+$, $(SO_4)^{2-}$

Multiple subscripts Multiple subscripts that apply simultaneously to a single host symbol would logically appear at the same level vertically. Potential confusion can be prevented by careful spacing of the collaborating elements, or through intervening commas or judicious use of parentheses.

Indices to indices If an index associated with a symbol must be modified by an index of its own, the natural result is a three-level expression, which would almost certainly require custom line spacing for surrounding text. "Secondary indices" of this kind would also mandate use of yet a smaller type size. Formulations of this sort quickly become awkward and difficult to interpret, however, so in the interest of clarity and legibility they should be avoided whenever possible. Thus,

Ex 18–25a $x_{n_{\max}}$

is better replaced by

b $x_{n(\max)}$ or $x_{n_{\max}}$

in which the secondary subscript "max" is set off in parentheses, or else shown in smaller type but raised to share a common baseline with the principal subscript.

The currently favored practice for typesetting chemical expressions is to consign all formulas to the normal text baseline, even ones that are intended to function as indices; in other words, the arrangement

Ex 18–26a $c(H_2SO_4)$

is preferred over

b $c_{H_2SO_4}$

Mathematical expressions often call for letters from the Greek alphabet, or other characters that might be described as "special symbols" (to serve as "operators", for example). If you find you need a symbol that unfortunately is not at your disposal, you may be compelled to create it by hand, preferably with help from an appropriate template. This may in turn oblige you to prepare a photocopy adjusted to the proper scale for eventual pasting into your document (cf. "Special Symbols" in Unit 19). A more sophisticated approach would involve scanning the symbol and "pasting" it electronically.

C 18–1   Is it fundamentally correct to claim that equations presented in a thesis must be italicized? If not, suggest a more accurate statement.

C 18–2   Should unit symbols be set in italic or roman type?

C 18–3   When is it legitimate to make use of a symbol ("abbreviation") like "cm", and when must one instead spell out the appropriate word (e.g., "centimeter")?

C 18–4   What distinguishing feature is usually associated with a "special function" in a typeset equation?

C 18–5   Numbers like –3, 12, or even e are routinely set in roman type. Under what special circumstances should (or might) a number be italicized?

C 18–6   What is the proper professional treatment of the symbols for vectors and matrices?

C 18–7   Identify a characteristic that the expression for the differential operator has in common with the symbol for "parts per million". In what ways does usage with respect to these symbols differ?

C 18–8   How should the size of type used for subscripts and superscripts relate to that for the corresponding host symbols?

C 18–9   Are symbols representing natural constants to be italicized?

C 18–10   Is it appropriate to set a numerical value directly adjacent to the associated unit symbol, as in

12m or 9.1mmol/kg?

How is the percent sign treated in this respect?

C 18–11   What notational problems do you detect in the following set of expressions?

12 mol/L   12 mol : L   12 mol $\cdot$ L^{-1} 12 $\cdot$ mol/L

C 18–12   Comment on the expressions shown below, and correct them as necessary:

2 dm   12 mµL   0.07 mmol   2450 mm   895 hPa   12 800 nA

C 18–13   Under what circumstances is it advisable to prepare a "List of Symbols" for a thesis, and where might such a list be placed?

C 18–14   Make whatever corrections you feel are appropriate in the following, which might have been taken from a typical "List of Symbols".

a_i impact strength, kJ/m^2
δ_0 = bulk density (g/cm^3)
H   spherical indentation hardness (in N/mm^2)
P   =   mean surface pressure [N mm^{-2}]
P_{WD} permeation coefficient for steam in g/cm $\cdot$ h $\cdot$ mbar

19 Mathematical Expressions and Equations

- This unit focuses on formatting issues related to mathematical expressions and equations, but it deals also with how one introduces such entities into a text document.

- The objective is to provide you with the knowledge required to handle in a professional way such "math" as you may need in your thesis.

Q 19–1 How should a minus sign be represented?

Q 19–2 In general, should equations be left-justified, right-justified, or centered?

Q 19–3 What role is played by spaces in a mathematical expression?

Q 19–4 To what extent is free space to be left beneath an equation?

Q 19–5 Is it necessary that all equations present in a document be numbered? And where would such equation numbers belong?

Q 19–6 What considerations arise when an equation proves too long to fit on a single line?

Q 19–7 Should punctuation appear at the end of an equation?

6.5

"Formulas" In a scientific context the word "formula" can refer to an expression based on either mathematical or chemical information. In this unit we will limit our use of the terms "formula", "expression" and "equation" almost exclusively to the former, reserving "structural formula" and "reaction equation" (or "reaction scheme") for presentations of the latter type.*)

Typographic considerations From a typographic perspective, (mathematical) expressions and equations can be regarded as especially complex examples of text. Their proper communication requires the "typesetting author" to be conversant with quite a

---

* We elected with this book to forego extensive treatment of chemical structural formulas.

number of technical rules, irrespective of the way a particular document is to be prepared (i.e., solely with the basic tools supplied by a word processor or with help from special mathematics software).

Expressions

The "text" in an expression of this kind consists largely of symbols of various sorts, where each symbol has a very specific meaning and plays a unique, essential role. Among the symbols one encounters in such a context (see also Unit 18) are lower- and upper-case letters of the Latin alphabet (a, b, A, B, etc.), Greek letters (α, β, γ, etc.), so-called operators, certain "special symbols" (including $\varnothing$, $\oplus$, or $\rightarrow$), and of course numbers. Meeting the inherent typesetting demands requires going well beyond the tasks normally assigned to a computer keyboard, although a few of the important unusual characters can be summoned by appropriate combinations of keys. The author facing the challenge of preparing mathematical copy should begin by becoming well acquainted with at least one of the special type fonts developed expressly for this purpose, the most familiar of which bears the *name* "Symbol" and is a standard feature on almost all PCs.

Equations

Preparing a specific equation in most cases entails systematically combining several relatively simple expressions, where the most basic arithmetic symbols are employed as connecting links:

Ex 19–1

$+ \quad - \quad \cdot \quad / \quad =$

Symbols for equivalence

One of these symbols—the sign of equality (or "equals sign", =)—could in fact be regarded as a source of the word "equation" itself. The slightly more complex symbol "$\equiv$" was once used to express an ostensibly more restrictive relationship—"identically equal to"—but the practice has been largely abandoned. The symbol itself continues to have a place in the special field of number theory, however, where it serves as a way of indicating congruence. The declaration that two things are equal *by definition* can if necessary be communicated through the variant "$=_{\mathrm{def}}$".

Non-equivalence

The most widely sanctioned symbol for *non equivalence* is "$\neq$". This could in principle be cobbled together by superimposing over the equals sign a diagonal line "╱" (but *not* the solidus "/", also known as a slash, which is set at an inappropriate angle).

$\leq, \geq$

The sign for "equal to or less than" (alternatively: "at *most* equal to") is "$\leq$", with "$\geq$" as its counterpart in the contrary sense ("equal to or greater than", "at *least* equal to").

Symbols for relationships that fall short of being "mathematical" in the strictest sense of the word include:

$\approx$ "equals approximately"
$\ll$ "is small relative to"
$\gg$ "is large relative to"
$\hat{=}$ "corresponds to"

Proportionality
One occasionally sees a "tilde" (~) employed in the sense of "equals approximately", but this mark should in fact be restricted to the very different relationship "is proportional to".

Addition, subtraction
Like the equals sign, the "plus sign" (+) is a mathematical symbol that also happens to have a place of its own on the standard typewriter or computer keyboard, but the key assumed by many to produce a "minus sign" (−) actually corresponds to the much shorter hyphen, "-", which makes a distinctly unsatisfactory substitute. Properly simulating a "minus sign" requires that one instead access the elongated dash many associate most closely with expressing a *range* (the "en-dash"; e.g. 2004–2006), and generating this in turn ordinarily involves knowing and then taking advantage of a special keystroke combination. The importance of the distinction should be clearly apparent from the illustrations in Ex 19–2:

Ex 19–2 $x - 4$ or Cl⁻ (*not* x - 4 or Cl⁻)

Multiplication
Under most circumstances the preferred rendition of a multiplication sign is a raised dot ($\cdot$), present for example in the PC font "Symbol". A perfectly acceptable (albeit less convenient) alternative would be an ordinary period (perhaps altered in size) raised to an appropriate level above the text baseline.

In many cases, multiplication can be efficiently indicated without recourse to any overt sign whatsoever, since scientists assume that when two or more quantity symbols are set close together they share a multiplicative relationship. For example, the combination

Ex 19–3 RT

would be universally interpreted as signifying multiplication of a particular thermodynamic temperature T by the ideal gas constant R.

Product formation (multiplication) can actually be expressed in several ways that are rather closely related; e.g.,

Ex 19–4 ab $a\,b$ $a \cdot b$
         $3b$ $3\,b$ $3 \cdot b$ (but *only* $3 \cdot 12$)

where the second expression in each of the above lines contains a "nonbreaking space". Whichever approach you take, it is wise to be generally consistent.

Also available for use in this context is the "multiplication cross" ($\times$). This character should whenever possible be extracted directly from a technical font like Symbol, *not* "approximated" by a lower-case letter "x". The same cross is called upon sometimes for format specifications; e.g.,

Ex 19–5   24 mm × 36 mm

The cross is an especially wise choice as a multiplication operator when there are decimal points nearby, since their presence could cause confusion; e.g.,

Ex 19–6 6.023×10^{23} (in preference to $6.023 \cdot 10^{23}$)

Vectors   When working with *vectors* you may find it necessary to distinguish between multiplications of two types: the kind leading to a "scalar product", as in

Ex 19–7   *ab*   or   *a* · *b*

and another requiring the presence of a multiplication cross to specify a true "vector product", also known (appropriately!) as a "cross product"; e.g.,

Ex 19–8
$$\begin{pmatrix} e_1 & e_2 & e_3 \\ a_1 & a_2 & a_3 \\ b_1 & b_2 & b_3 \end{pmatrix}$$

Division   The act of division, too, can be expressed in various ways, including:

Ex 19–9 a/b ab^{-1}

In a scientific setting one should never represent quotient formation with ":" or "÷". When there is need to communicate a complex expression in which division plays a significant part, maximum clarity usually results from the classic notation that relies on a horizontal fraction line of appropriate dimensions, as opposed to use of either slashes or the "trick" of negative exponents. Ex 19–10 offers a fairly dramatic illustration of the point, through three strikingly different depictions of the same information:

Ex 19–10
$$z = [1/(xy) + 1/(\ln x)]/[x/(x + y) + y/(x - y)]$$
$$z = [x^{-1} \cdot y^{-1} + (\ln x)^{-1}] \cdot [x \cdot (x + y)^{-1} + y \cdot (x - y)^{-1}]^{-1}$$
$$z = \frac{\dfrac{1}{x\,y} + \dfrac{1}{\ln x}}{\dfrac{x}{x + y} + \dfrac{y}{x + y}}$$

Unfortunately, the extra legibility achieved with the third equation above may come at the expense of considerable typographic effort, probably the main reason why formulations with negative exponents are so popular among scientists and engineers, and are often resorted to as a kind of "second-best solution".*)

---

* For footnote see page 146.

When division is indicated with a slash, avoiding ambiguity frequently makes it necessary for the expression to the right of the slash to be set in parentheses. Consider, for example,

Ex 19–11a $a/b/c$ $\ln x/3$ $p \cdot V/R \cdot T$ $mL/m^2 h$

all of which are precariously subject to more than one interpretation; i.e.,

b $a/b/c$: $a/(b/c) = (ac)/b$ or $(a/b)/c = a/(bc)$
$\ln x/3$: $\ln (x/3)$ or $(\ln x)/3$
$p \cdot V/R \cdot T$: $p \cdot V/(R \cdot T)$ or $(p \cdot V/R) \cdot T = p \cdot V \cdot T/R$
$mL/m^2 h$: $mL/(m^2 h)$ or $mL\ h/m^2$

Spacing   *Spacing* with respect to symbols is always an important consideration in mathematical typography. For example, single blank spaces (which in a professional rendition would be both narrow and of fixed width) should be inserted before and also after such symbols as $+ , - , = , \cdot , < , >$ (among many others); e.g.,

Ex 19–12a $z = a + 2b - c$ $x < 12.5$

A little free space is also necessary before and after symbols representing certain functions; e.g.,

b $2 \sin x$ $a \tan 3b$

In some formulas the presence or absence of extra space can have an important influence on interpretation. Thus, spacing is regularly employed to suggest *grouping* with respect to elements that are in some way related. This becomes especially critical in expressions that contain function symbols; e.g.,

Ex 19–13a $a \sin \omega t\, e^x$

A reader could be expected to recognize here that "ωt" and "e^x" represent two distinct entities. The resulting message is thus very different from that embodied in

b $a \sin \omega\, te^x$

All risk of misinterpretation can of course be avoided by the judicious disposition of parentheses:

c $(a \sin \omega t)\, e^x$

---

* You should be aware, however, that powerful assistance exists in the form of auxiliary software specifically designed to facilitate preparation of complex mathematical expressions. These helpers are referred to as "equation editors", and they can provide remarkably good approximations of professional mathematical typesetting, which one can then essentially "paste" into the corresponding text file exactly where they are needed. Nearly all the equations in this book, for example, were prepared with just such software: Microsoft's EQUATION EDITOR. The latter is supplied as a module accompanying Microsoft OFFICE, but is in fact only a subset of a more versatile "stand-alone" commercial package from Design Science called MATHTYPE. A similar program is InfoLogic's MATHMAGIC. Both are cross-platform (Windows, Macintosh) applications. If you anticipate needing quite a few equations of even moderate complexity we strongly recommend you consider availing yourself of one of these valuable tools.

"Fences" Parentheses, together with other types of "fences" (such as [ ] and { }), perform a critical service in mathematical expressions by strongly reinforcing inherent structure, making clear, for example, the amount of territory to which a particular function applies. Sometimes multiple "nested" fences are required, and if so they should be erected in the following pattern:

Ex 19–14 $\{[(\ )]\}$

i.e., parentheses inside square brackets, which are in turn treated as subordinate to "curly brackets" (braces). Fence symbols of every type should be expanded vertically as much as necessary to ensure that they actually embrace all the material within them.

Miscellaneous symbols Integral, summation, and product symbols are similarly subject to scaling so that they at least approach the optimum height for expressions to which they apply, as for example

Ex 19–15 $\displaystyle\sum_{i=1}^{n} f(x_i) \quad \sum_{i=1}^{\infty} \frac{f(x_i)}{g(x_i)}$

Other symbols frequently required in mathematical expressions include the square-root sign (also with dimensions that should be treated as flexible) and the familiar sign for infinity (∞), together with a wide assortment of special characters native to specialized branches of math, like set theory, as well as the rich world of physics; e.g.,

Ex 19–16 $\Delta \ \partial \ \nabla \ \uparrow \ \downarrow \ \varnothing \ \in \ \cup \ \cap \ \rightarrow \ \Leftarrow \ \Uparrow \ \wedge \ \Omega \ \perp$

As suggested earlier, many of the symbols most often in demand for equations are included in the type font "Symbol", which is available (or at least *should* be available) on almost every PC. This in turn ensures high quality output of equations at any desired scale with nothing more complex than an ordinary PC printer.

In the unlikely event you do not have at your disposal a powerful word processor for the flexible manipulation of text elements, or you lack access to one or more essential symbols, a cumbersome but effective workaround with respect to mathematical material is preparation by hand of an appropriate-sized rendition in a free space left specifically for it. You will almost certainly achieve better results if you succeed in locating suitable templates for the purpose. Another possibility is to search through the inventory of a drafting supplies dealer in the hope of finding what you need in the form of adhesive transfer characters, though these have become increasingly rare as more and more technical drawing is carried out on computers. If you do find yourself resorting to traditional templates and drafting pens we advise that you prepare your drawings to a 3.5- or 5-mm standard, and then reduce them photographically or with a photocopier to the correct scale for pasting into the master document.

Free-standing expressions With the exception of very small expressions like

Ex 19–17 $x = 3$ or $T = 298$ K

mathematical elements in formal printed works are usually made into free-standing elements, not incorporated into running text. In other words, each is assigned its own separate line, where it is generally indented—and to an extent significantly greater than typical paragraph indentation. Space equivalent to one-half a text line (or perhaps even a full line) should separate an equation from preceding and following text. For example,

Ex 19–18   Text text text text text text text text text text text text text text text text text text text text text text text text text text text text text text text text text text text text text text text text text text text text text

Formula

text text text text text text text text text text text text text …

Equation blocks If several such equations appear together—one above another, creating a free-standing equation *block*—they should be mutually spaced consistent with the surrounding text, at least so long as each can be accommodated on a single line. With expressions more demanding in the vertical dimension—such as ones involving complex fractions, roots, or summations—the separation between equations, and between text and equations, must of course be adjusted accordingly; e.g.,

Ex 19–19 $y = \sin(x + \pi) + \cos(x - \pi)$

$$y = \frac{x^2 \cdot (1 - x) \cdot \ln(2x - 1)}{(x - 2) \cdot (x + 2)}$$

$$y = \frac{\dfrac{x}{x-2} \cdot \ln(2x - 1 + e^x)}{(x - 2) \cdot (x + 2) \cdot \dfrac{\sin x}{x + 0.1}}$$

With sequences of equations like those used for depicting a mathematical derivation, greater clarity can be achieved by aligning all the various elements on the basis of their equals signs rather than indenting uniformly.

Equation numbers It can be quite advantageous in a lengthy document to distinguish the equations present by *number*, the appropriate numbers being set in parentheses flush against the right margin, at the same level as the formula axis. For example,

Ex 19–20 $f(x) = ax^2 + bx + c$ (12)

An important advantage of numbered equations is that they are easily referred to from within the text. At the very least you should number all equations to which you will in fact make reference.

Placement Equations can be introduced into the body of a document in two different ways. One possibility is simply setting them immediately after the individual paragraphs to which they most clearly relate, prior to the start of the next paragraph (in essence creating separate "formula paragraphs"); e.g.,

Ex 19–21 Text text text text text text text text text text text text text text text text text text text text text text text text text text text text text text text

$$z = f[x, \varphi(u,v)]$$

Text text text text text text text text text text text text text ...

The alternative is treatment as in the following example:

Ex 19–22 It stands to reason, therefore, that

$$dc_A/dt = -kt$$

which in turn after integration under the initial conditions specified in Eq. (4)

$$c_A = c_A^0 \, e^{-kt}$$

leads to ...

Embedded In a case like this the paragraphs of text no longer enjoy a separate exist-
equations ence, and formal sentence structure may declare that punctuation of some sort (a comma or a period) should appear at the end of a mathematical expression. Such punctuation is often omitted, however, since it tends to look "lost" in conjunction with a complicated collection of symbols, and could even be mistakenly perceived as part of the math.

"Breaking" an If for some reason you decide not to make a lengthy expression free-
expression standing, it is important to make absolutely sure it cannot become a victim of "automatic" (random) line feeds. An outcome like the following, for example, is totally unacceptable:

Ex 19–23 Text text text text text text text text text text text text text text text text $(x_1 - y_1)/$ $(x_1^2 - y_1^2)$ text text text text text text ....

When complete rendition of an equation requires more than a single line, the expression should be divided: immediately before a convenient plus or minus sign, but *not* one that falls within the bounds of a set of fences; e.g.,

Ex 19–24 $c^4(H_3O^+) + c^3(H_3O^+) \cdot K_{S1} + c^2(H_3O^+) \cdot (K_{S1} K_{S2} - K_w - CK_{S1})$
$\quad - c(H_3O^+) \cdot (2 \, CK_{S1}K_{S2} - K_{S1}K_w) - K_{S1}K_{S2}K_w = 0$

Actually, the best place to "break" a lengthy equation is right ahead of an equals sign. As previously noted, you should also try to stack groups of equations such that equals signs are aligned; e.g.,

Ex 19–25 $p_0 = k_1 h^2 N^{5/3} m^{-1}$
$= k_2 N e_F$
$= k_2 U_0 V$

C 19–1   What have you learned about representing the minus sign?

C 19–2   In what various ways might fractions be expressed in a mathematical formula?

C 19–3   Correct formal errors you detect in the following text excerpts:

a   … produces consistent with the relationship

$$k_1 = k_n c_2 n^{-1} + k_0 \tag{22}$$

and in this special case of n>3:

$$k_1 = k_2 c_1 + k_0 \ (i=1,\ldots n) \tag{23}$$

b $y = 3{\cdot}(x - 2{\cdot}[x + 3]{\cdot}\{x^2 - 2x + 14\})$

c $\int\limits_0^\infty \dfrac{f(x)}{g(x)} \qquad \overset{n-1}{\underset{i=1}{\Sigma}} f(x_i)$

C 19–4   Pick out formal errors in the passages that follow:

a   … this odd result obtained by introducing the standard values of $x_{AWS} = 125$ mbar and $y_a = 12$ mmol/l, $y_b = 0.01$ mmol/L and $y_c = 10$ mmol/L …

b   … can be accurately described with the aid of the three equations

$f_1(x) = a \cdot x^2 + b \cdot x + c, \quad (5)$
$f_2(x) = (a + 1) \cdot \ln(1 - x), \quad (6)$
$f_3(x) = b \cdot \exp(-x^2 + 2). \quad (7)$

The obvious conclusion, especially with respect to …

20 Tables

● In this unit you will learn about the proper use and design of tables.

■ The information provided will position you to present experimental data and other information effectively in tabular form.

Q 20–1 When should a table be "inverted"?

Q 20–2 What is meant by "anchoring" a table in the text?

Q 20–3 In what ways does a list differ from a table?

Q 20–4 How should quantities and units be expressed in a table heading?

Q 20–5 To what extent should tables resemble body text (in terms of line spacing and type size, for example)?

Q 20–6 How might one draw attention to specific values, rows, columns, or regions in a table?

Q 20–7 Should numerical data in a table be left-justified, right-justified, centered, or aligned according to the decimal point?

 | 8.1–8.4

Definition   Tables are presentations in which verbal, numeric, or graphic information has been arranged in a systematic way in rows and columns. In a lengthy document like a thesis, tables encourage the reader to pause and think about their content, and they provide the author an opportunity to present information in a concentrated form. Rather like figures (cf. Unit 21), tables have the power to capture a reader's attention. Also like the former, if properly designed they are fully capable of standing on their own, independent of the host document's text.

Table captions   A table is nearly always introduced with a caption. This usually commences with an identifying number, to be followed by a formal title that announces the table's content.

Table numbers are traditionally declared in the form "Table X", separated from the accompanying title by a period. (In rare instances a colon appears in this context rather than a period.) In printed works this very first element is generally set in boldface type, less often in italics. "Double numbers" (e.g., "Table 2–5.") are desirable in the case of long documents divided into chapters.

Ex 20–1   **Table 4.** Enol content ω (in %) ⎤———— Table caption
for **1** and **2** in various solvents. ⎦

Solvent	**1**	**2**
H_2O	0.4	1.8
C_2H_5OH	12.0	12.5
C_6H_6	16.2	16.0
CS_2	32	34.8

"Rules"

The reason tables are assigned numbers is to facilitate making reference to them individually from within the text. Indeed, it is considered essential that *every* numbered table be referred to—by number!— at least once somewhere in the document, a practice known as "anchoring" the table. For example:

Ex 20–2   … these values (see Table 6) show that …
… is known (cf. the third column in Table 4–5) …
… are collected in Table 12.

Table titles   The title briefly characterizes a table's content, often identifying the source or purpose of the collection as well. Normally, such titles are formulated without recourse to verbs; i.e., essential terminology is strung together with conjunctions, articles, and prepositions (cf. the "for", "and", and "in" in the title accompanying Ex 20–1). Try if at all possible to accomplish the objective with a *single* "sentence", which should end with a period (a detail that is often overlooked!).

Commentary   The official title of a table is frequently followed by a certain amount of explanation. As in the case of the analogous "legends" attached to figure captions (cf. Unit 21), one should set such an elaboration apart visually from the title itself, not only by the obligatory period, but also by a prominent dash:

Ex 20–3   **Table 2–5.** Elemental composition (in terms of amount of substance) of dried plant material relative to phosphorus content. — Elements labeled with an asterisk are considered primary nutritional constituents.

Major components		Trace materials	
Element	Rel. content	Element	Rel. Content
H	470	Cl	0.66
C	250	S	0.53
O	170	Si	0.31
N*	9.1	Na	0.20
K*	3.5	Fe	0.12

Comments providing details related to a table are advantageously consigned to *table footnotes* (cf. also Unit 17). These are still considered part of the corresponding table, but appear immediately below the line marking the end of table content. Footnotes can be associated with any part of a table, including the title. Responsibilities commonly delegated to footnotes include clarifying unfamiliar abbreviations or acronyms, indicating exceptional conditions that affect specific data entries, quantifying standard deviations, and the like. Every footnote, like the title, should end with a period. The markers for signaling table footnotes are ordinarily lower-case (roman) letters followed by close-parenthesis signs, where the latter are as usual omitted at the bottom of the table, in front of the notes themselves. *Numeric* markers should be avoided to eliminate any chance for confusion with (numeric) table content.

Ex 20–4   **Table 3.** Chlorinated benzene derivatives in exhaust gases released during polyethylene combustion in the presence of **1**.

Peak No.	Compound	Concentration[a] (in μg/g)
1	Chlorobenzene	10
2	1,3-Dichlorobenzene	1.0
3	2,3,7,8-TCDD[b]	0.008
...		

[a] Based on the polyethylene charge.

[b] 2,3,7,8-Tetrachlorodibenzodioxin ("Seveso dioxin").

Occasionally symbols like *, †, or ‡ perform the alerting function.

Explanatory material that applies to a table as a whole sometimes is conveyed not as an extension of the table title, but underneath the content, immediately preceding any footnotes. No special identifying markers are required with addenda of this sort.

Table heading

The first line (or row) within the actual table functions as a heading, providing text that clearly identifies the nature of ensuing entries in the various columns. Each column must be supplied with a label of its own (known as the "column heading").

Formal aspects of heading structure

In order to establish a clear visual separation between a table heading and subsequent tabulated data, the heading should be flanked by a pair of horizontal lines or "rules" (see Ex 20–1). Heading entries must of course be very brief, given the limited space available, although they are allowed to extend over multiple lines. As one example of how the special "Table" function (see below) supplied with most word processors greatly simplifies the process of table construction: whenever a column heading threatens to spill over onto a second line, such a program automatically expands the vertical space allocated to headings in general, across the entire width of the table—and to precisely the extent necessary.

Sometimes the nature of material to be presented in a table is so straightforward, based on the title, that one can dispense with a formal heading, as illustrated in Ex 20–5.

Ex 20–5

Table 3–4. Selected physical characteristics of the element fluorine.

Boiling point	85.0 K
Critical pressure	52.2 bar
Critical volume	$1.74 \cdot 10^{-3}$ m^3/kg
Density at 77.8 K	1562 kg/m^3

Quantity symbols

Table headings in scientific publications frequently rely heavily on symbols rather than words, especially quantity symbols (which, as pointed out in Unit 18, are always to be set in italics).

Ex 20–6

c_0 (in mmol L^{-1})	E (in g)	ε_1 (in L mol^{-1} mm^{-1})
145.2	112.3	2460
...		

Column widths

Widths assigned to individual columns should be a function of the specific requirements of column content. All column entries, together with the corresponding heading, normally should be left-justified. This has the advantage that the resulting perfect vertical alignment helps clearly delineate column boundaries. If necessary, alignment of an analogous sort can be achieved (albeit awkwardly) within the ordinary text mode of a word processor through judicious application of tabs and the tab key, but this is a place where the program's dedicated "Table" function reveals itself as a true godsend. Thus, one is able to start with a custom template that amounts essentially to a blank table, containing exactly the right number of columns and rows. All that remains for the author to do is supply content. Dimensions for the various columns can be adjusted quite easily at any time so as to satisfy one's needs precisely (most of which will become obvious as part of the data-entry process). Should your plans for some reason change, rows and/or columns can be added (or deleted) at will. You also enjoy access to facilities for highlighting individual rows, columns, or even cells ("data compartments") through either borders or background colors (including gray shades). Selected parts of a table are subject as well to extensive custom formatting. By all means, when the time comes to prepare your first tables, set aside a few minutes for experimenting with the available possibilities, seeing for yourself the various effects that can be achieved.

As noted above, at least some of the column headings you create are likely to involve quantities, ordinarily expressed symbolically. A heading also serves as the perfect place to identify *units* applicable to your data, and if necessary to report order-of-magnitude (power-of-ten) scaling factors the reader needs to take into account. The former eliminates the need to make units explicit in conjunction with every single data entry in those cases where all labels would be the same (which is not always true, however, as seen from the table in Ex 20–5). This more efficient arrangement is illustrated by

Ex 20–7

$c(HA)$	$m(CaCl_2)$	p	U	L_m
mol L^{-1}	kg	kPa	10^6 t	S m^2 mol^{-1}

...

Units   Although one occasionally sees it done, you should resist any urge to enclose unit symbols in square brackets within a heading. Setting them in *parentheses*, under the corresponding quantity symbols, is an appropriate expedient, however, since the units can then be interpreted as "explanatory text". Units cited in this way are sometimes preceded by the word "in" for further clarification (cf. Ex 20–6).

Ex 20–8

V	p	T
L	Pa	°C

V	p	T
(L)	(Pa)	(°C)

Powers of ten   Table-heading information related to order-of-magnitude (power-of-ten) factors that affect tabulated data is notorious for being a source of confusion, a problem that plagues axis labels for graphs as well (see Unit 21). More precisely, too many authors fail to make sufficiently clear whether reported data (including the stated units) are to be viewed as 10^x fold values of experimentally determined quantities, or if displayed information should instead be *multiplied* by 10^x. (For a very general review of the overall subject of quantities and units, by the way, see Unit 18.) Consider the following samples of notations one sometimes encounters:

Ex 20–9

$10^{10}\, r$
m

or

r
10^{-10} m

or

r
10^{-10} m

Quotient notation   At least some readers would have trouble interpreting the first two illustrations in Ex 20–9, but the third is better, and warrants special attention due to its "scientific" nature, broad applicability, and lack of ambiguity. Thus, a "quantity" is by definition the product of a numerical value and a unit (cf. p. 134), so *dividing* a quantity by its units, as here, must logically produce a pure numerical value, and it is precisely these numerical values that one ordinarily finds collected in tables. Thus, with the third heading option above, a numerical entry of "1.54" would (according to the column heading) constitute one specific instance of $r/(10^{-10}$ m); in other words, in this context $r = 1.54 \cdot 10^{-10}$ m. That said, however, it is still better to make creative use of unit prefixes so as to avoid any need for burdening the reader with power-of-ten factors.

Structured headings   Structure can be conferred on a complex table heading in an effective way by introducing a few horizontal and vertical lines. Vertical lines in this case can be regarded as extensions of column boundaries, leading to "compartmentalization" of the heading, whereas horizontal lines have a unifying effect, creating "table headings within table headings". For example:

Ex 20–10a   **Table 12.** Relative toxicities of selected poisons.

Substance	Molar mass	Minimum lethal dose	
	in g/mol	in mol/kg	in mg/kg
Botulinus toxin	$9 \cdot 10^5$	$3.3 \cdot 10^{-17}$	0.00003
Tetanus toxin	$1 \cdot 10^5$	$1.0 \cdot 10^{-15}$	0.0001
TCDD	322	$3.1 \cdot 10^{-9}$	1
Sixitoxin	372	$2.4 \cdot 10^{-8}$	9

As a consequence of a table's inherent two-dimensionality, each entry lays claim to its own distinct space (or "cell"), which one can in turn specify through a pair of coordinates. This feature is exploited especially heavily by data management tools associated with spreadsheet software.

The column situated at the far left of a table—known in English as the "stub"—often has special significance, in that the various entries may serve to define the series of cells immediately to the right (i.e., the other components of a particular row) in much the same way the parts of a table heading label various columns. In tables consisting largely of numerical information, the leftmost column frequently lists the several "independent quantities" studied. Alternatively, it could detail various sets of experimental conditions, or simply supply sample numbers—again, information unique to values present in particular rows with regard to the *dependent* quantities identified in the different column headings.

In view of their two-dimensionality, tables are sometimes described as digital counterparts of graphs designed to give visual expression to a relationship between two variables. In publications it is ordinarily not permissible to provide equivalent data in both analog and digital fashion, largely in the interest of efficient utilization of space. Results are instead conveyed *either* through a table *or* with a graph. Theses are less restrictive in this respect, however, and dual presentation may even be encouraged. It is worth noting, by the way, that much computer software used for data manipulation is capable of generating professional-looking output of both types from a single set of data.

Sometimes it becomes necessary to commit more than one column to characterization of a particular "independent quantity". Thus, a first column might be dedicated to reporting the overall nature of material under investigation, with the next column used to identify specific samples. Only in a third column would true data make an appearance.

Boundary lines   Some authors make a habit of building "walls" around the cells that make
up their tables, in the form of lines that establish firm row and column
boundaries. The software employed for table creation can thus be instructed
to add substance to the grid lines (initially present only as "metainforma-
tion") faintly revealed on the screen during table construction. An abun-
dance of lines within a table is actually more likely to engender confusion
than insight, however, and what results may conjure up images of bars
installed across a window: surely not an aesthetic response one would wish
to encourage. Note the contrasting impressions made, for example, by the
two treatments illustrated in Ex 20–10a and b.

Ex 20–10b   **Table 12.** Relative toxicities of selected poisons.

Substance	Molar mass	Minimum lethal dose	
	in g/mol	in mol/kg	in mg/kg
Botulinus toxin	$9 \cdot 10^5$	$3.3 \cdot 10^{-17}$	0.00003
Tetanus toxin	$1 \cdot 10^5$	$1.0 \cdot 10^{-15}$	0.0001
TCDD	322	$3.1 \cdot 10^{-9}$	1
Sixitoxin	372	$2.4 \cdot 10^{-8}$	9

Vertical lines in a table tend to be especially unsightly. Rather than adding lines to emphasize column boundaries it is much better in general to
be content with the plain-paper stripes already present.

"Empty" cells: There is, by the way, no rule insisting that every cell in a table must have
$0, -, \ldots$ something in it. For example, if for some reason there is no way you can
associate a particular cell with data, why clutter it up with a (superfluous)
dash ("–")? Embellishment of this sort is especially unwelcome if you fail
to provide a footnote to explain how such a mark is supposed to be interpreted. (Possibilities could include "not applicable", "not determined", and
"undefined".) It is sometimes recommended that one indicate "not determined" by "…" and "undefined" (or "inapplicable") by an empty cell. This
avoids the dashes that might in any case be confused with minus signs,
while leaving open the option of deploying a true minus sign in situations
where its presence could actually make sense (as when a test was simply
not carried out).

Tables in journal articles and books are almost always set in type smaller
than that employed for body text, in part to facilitate the accommodation
of tabulated material. Smaller lettering (e.g., 10 pt in the case of 12-pt body
text) of course permits more information to be incorporated into each cell,
with less need for continuation lines. If cell content still must be distrib-

uted over two or more lines, though, feel free to introduce breaks between words or, if necessary between syllables (cf. Table 20–11). The possible alternative of "inventing" novel, space-saving abbreviations should be categorically rejected.

Ex 20–11   **Table 4**. ASTM Specifications for propanols.

	1-Propanol	2-Propanol	Specification method
ASTM standard	D 3622-90	D 70-90	
Color, Pt-Co (ASTM)	10	10	D 1209
Distillation range 101.3 kPa	distill within a 2 °C range that includes 97.2 °C	distill within a 1.5 °C range that includes 82 3 °C	D 1078
Nonvolatile matter mg/1000 mL	5	5	D 1353
...			

Should you be forced for whatever reason to create a table manually on the basis of ordinary type, one "gimmick" worth considering is narrower line spacing relative to body text as a sly way of simulating smaller type.

Proper typesetting of columns   For the reader's convenience with respect to interpretation, you should in most cases arrange columns of numbers such that decimal points (or ± signs) are uniformly aligned. This may well produce columns that are "ragged" along the edges, but that can be advantageous, because it makes it easy to spot unusually large and/or small numerical entries.

Ex 20–12

12.4	12.4	12.4
8.7	8.7	8.7
1120	1120	1120
0.85	85	0.85

On the other hand, if it would be pointless to compare various values within a column (because they bear no direct relationship to one another), it does *not* make sense to practice decimal-point lignment (cf. Ex 20–5).

≤ , ≥   Table organization need not be a function of the heading alone: sometimes the left-most column (the "stub") participates as well. It often helps, for instance, to group the entries in this first column, and/or introduce occasional "headlines" or blank rows. Consider, for example, the effect of reconfiguring Ex 20–13a

Ex 20–13a

A (USA)	000	000	000
A (Europe)	000	000	000
A (Japan	000	000	000
B (USA)	000	000	000
B (Europe)	000	000	000
B (Japan)	000	000	000

...

into a pattern like

b

USA

A	000	000	000
B	000	000	000
C	000	000	000

Europe

A	000	000	000
B	000	000	000
C	000	000	000

Japan

A	000	000	000
B	000	000	000
C	000	000	000

If you have a sense that one of your tables is perhaps too long and narrow, try inverting it, as in the following example:

Ex 20–14a

No.	p(in kPa)
1	21.0
2	31.4
3	45.6
4	54.8
5	62.8
6	69.2
7	72.6
...	

b

No.	1	2	3	4	5	6	7	8	9
p (in kPa)	21.0	31.4	45.6	54.8	62.8	69.2	72.6	78.4	81.1

Another way to deal with the problem of an overly long table is to "break" it into segments; e.g.,

c

No.	p(in kPa)	No.	p(in kPa)
1	21.0	5	62.8
2	31.4	6	69.2
3	45.6	7	72.6
4	54.8	8	78.4

Table size Try to avoid including tables in your thesis that are either too large or too small. One that is too small is not worth the effort invested in its creation, unless what it contains is extraordinarily important. In most cases, carefully organized body text would accomplish the same ends. On the other hand, the content of an exceedingly large table seldom entices the reader to pause and reflect, creating a risk that important information will be overlooked. A "comfortable" compromise typically occupies roughly half a page.

Condensing a table If you are afraid one of your tables could prove overwhelming, see if you can compress it by deleting the least important columns, or ones housing data that reveal little variation. Some of what you discard in the process might still be passed along to the reader through table footnotes. It could also be that numerical values reported in a particular column are in fact easily derived upon inspection from those in another column by straightforward calculation (e.g., multiplication by a conversion factor). As a last resort, consider dividing an "overweight" table into two smaller ones.

If none of the above strategies shows promise, the offending table should probably be consigned to an appendix, with references to it in the text as necessary. In the event such a table actually extends across multiple pages, be sure the table heading is repeated in its entirety at the top of each new page, possibly accompanied by a reminder like

Ex 20–15 **Table A.x** (continued)

Importance of clarity Take special pains to achieve optimum clarity in tabular presentations. Our earlier suggestion that numbers be aligned according to their decimal points is of course one step in that direction. Other useful expedients include placing adjacent to one another columns that are subject to comparison, organizing data according to a logical scheme (e.g., in order of ascending numerical values, for example) rather than by—essentially arbitrary—experiment numbers, and analogous treatment of tables that are structurally similar. Keep in mind also that the typical reader approaches table content in the same way he or she would a page in a newspaper: one column at a time, proceeding from left to right. This implies that comparisons will be made most easily among entries in a given column as opposed to ones sharing a common row.

Uniformity Be consistent also from table to table in your use of terminology, abbreviations, and symbols—and between tables and body text as well. In the same vein, use similar footnote markers in all your tables. As a rule, one should deploy letters a, b, c, ... sequentially for this purpose, introducing them systematically from the upper left to the lower right, just as one would in text.

One of the very last steps preparatory to printing a final copy of your thesis should be one further check to make sure all tables are correctly numbered, and all are also suitably anchored in the text.

Lists A slight digression seems in order before we conclude this unit. If a compilation consists of nothing more than an ordered identification of a set of words or phrases it is referred to not as a table but as a *list* (or listing), despite the fact that it could plausibly be described as a table of the most primitive kind. Whenever possible, lists incorporated into running text should be indented relative to their environment as a way of subtly calling attention to them; e.g.,

Ex 20–16 Multiple bonds between atoms are always shorter, for example, than corresponding single bonds, as illustrated by the various bonds between pairs of nitrogen atoms:

$N\equiv N$ (110 pm)
$N=N$ (125 pm)
$N–N$ (145 pm)

One explanation for this phenomenon starts from the fact that the covalent radius of an atom depends upon that element's ...

Especially if a list is composed of text passages, individual elements should be highlighted with "bullets" (•), or distinguished by introductory hyphens or dashes (e.g., - or –). For instance

Ex 20–17 ... and for this reason we felt it essential that our investigation take into account the three possible carriers

- water,
- air, and
- soil

since all three might contribute significantly ...

Neutral symbols like the bullet should give way to numbers or letters only if you propose to make reference within the text to specific points from the list; e.g.,

Ex 20–18

1.   a)
2. or b)
3.   c)
etc.

Note in this illustration that the *numbers* are shown followed by periods (for reading as "first", "second", etc.), whereas after *letters* there are close-parentheses signs (but one never uses both!).

C 20–1   Make appropriate corrections or improvements in the tables that follow.

a Table 1.

x	y
1	123
10	22
25	84
40	315
5	118

b Table 4-2. Time-dependence of the concentration c of the solution.

c	t
$5.052 \cdot 10^{-4}$ mol/l	42′
$1.526 \cdot 10^{-6}$ mol/l	0.5min
$7.015 \cdot 10^{-6}$ mol/l	2′
$1.033 \cdot 10^{-4}$ mol/l	19′30 sec
$4.728 \cdot 10^{-5}$ mol/l	15′
$0.261 \cdot 10^{-5}$ mol/l	12′30sec

c **Table 4-4**. Yields in the reaction of CH_3SO_2Cl with H_2NOH.

Temperature	% Yield
50°C	65%
60	68%
70	75%
90	92%
100	95%
110	95%
120	59%*

* partial decomposition was noted

d **Table 12**. Properties of several solvents.

	molar mass (in g/mol)	m.p. (in °C)	b.p.[a] (in °C)	d (in g/cm^3)
C_6H_6	78.12	5.5	80.1	0.87865
Phenol	94.11	43	181.75	1.0722
Toluol	92.15	-95	110.6	0.8669

[a] at 760 mbar

C 20–2 Critique the following table, including the way it is anchored in the associated text fragment.

… the amount of **13b** present, which was measured at two points in the smokestack (see Table 2–1), varied around the value 7 ppm, and was thus significantly below the 50-ppm limit prescribed by the federal government [12].

Table 2–1. Level of **13** in flue gas.

Measurement site	Time	Flue-gas temperature	Content	Flow rate
1	5 min	606°C	8 ppm	2.5 m³/min
1	10 min	604°C	6 ppm	2.5 m³/min
1	15 min	605°C	7 ppm	2.5 m³/min
1	20min	606°C	6 ppm	2.5 m³/min
1	30 min	604°C	7 ppm	2.5 m³/min
1	1 hr.	608°C	7 ppm	2.5 m³/min
2	30 min	606°C	7 ppm	3.0 m³/min
2	1 hr.	606°C	7 ppm	3.0 m³/min

C 20–3 Transform the following text passage into text supported by a table.

The earlier literature reports the following physical properties for compounds of the type $CH_3-SO_2-N(R^1)OR^2$: observed melting points for the compounds with $R^1 = R^2 = CH_3$ and $R^1 = R^2 = H$ are, respectively, 252 °C (with decomposition) and 172 °C; for $R^1 = H$, $R^2 = CH_3$ and $R^1 = CH_3$, $R^2 = H$, 132 °C and 145 °C; refractive indices with $R^1 = R^2 = CH_3$, $R^1 = H$, $R^2 = CH_3$, and $R^1 = CH_3$, $R^2 = H$ are 1.4352, 1.3528, and 1.4255 [3].

C 20–4 Incorporate the information in the following table into a suitable text passage.

Table 7. Intensity ratios for peaks at m/e = 44 and 46 in the mass spectrum of dinitrogen monoxide released in the decomposition of **3** in aqueous solution.

Experimental parameter	^{18}O-enriched water	Non-enriched water
I(46)/I(44)	0.0020 ± 0.0005	0.0025 ± 0.0012

C 20–5 Which of the following table headings would you consider appropriate to offer as models?

a

r in mm	I_0		
	in eV	in kcal	in kJ

b

r	I_0	I_0	I_0
[mm]	[eV]	[kcal]	[kJ]

c

r/mm	I_0/eV	I_0/kcal	I_0/kJ

d

r	mass	d	boiling point
mm	mg	g/cm^3	°C

C 20–6 Suggest ways the following table might be improved.

Experiment no.	$c(RSO_2^-)$ (in mol/L)	$10^3 k_1$ (in min^{-1})	
		Found	Calcd.
1	0.0125	29.0	31.1
2	0.025	14.2	13.9
3	0.05	6.9	5.8
4	0.1	2.58	2.14

21 Figures

- This unit discusses the design and preparation of copy for scientific and technical illustrations, its incorporation into effective figures, and the considerations involved in using, in a thesis, illustrative material taken from an external source.

- On the basis of this introduction, and given the necessary tools, you should find yourself able to prepare first-rate figures, and with little or no outside help.

Q 21–1 What is the distinction between a "qualitative" and a "quantitative" approach to displaying information graphically?

Q 21–2 How do line drawings differ from halftone illustrations?

Q 21–3 How can a "legend" (in the strictest sense of the term) contribute to a figure caption?

Q 21–4 If the author of a thesis wishes to reproduce a figure already published elsewhere (e.g., in a journal or a monograph), what formalities must be attended to?

Q 21–5 To what scale should one prepare a line drawing intended for subsequent use as a figure? What role can photoreduction play? And what guidelines exist with respect to size for the lettering in a figure?

Q 21–6 How thick should you make the lines for curves, axes, and other features in a graph you plan to reproduce in a thesis?

 7.1–7.5

21.1 Figure Captions, and Links Between Figures and Text

Figures, illustra-tions, pictures, and graphics
In the discussion that follows we regard as a "figure" any element within a document that cannot be generated readily with ordinary word process-ing tools—utilizing, apart from standard alphanumeric characters, only the small set of commonly available symbols. This obviously includes (among other things) technical drawings, photographs, and the various graphic approaches to displaying numerical data.—"Graphics" is in fact another term frequently employed in this same context, especially with respect to computer-generated materials.

The significance of figures
Figures sometimes function as "primary" information sources, in that they are the site of actual experimental results (e.g., a spectrum, or a photo-graph of something subject to investigation), or they interpret some method or technique in an especially definitive way, as with a detailed illustration of an experimental setup. More often, however, a figure represents a "sec-ondary" vehicle: a tool for enlivening communication. That is to say, most figures confront the reader with a "derivative" image, one prepared ex-pressly to assist the author in making a point, as for example a graph or a schematic diagram. Figures can enrich a thesis in many contexts, includ-ing in the "Results", "Discussion", and "Experimental" sections.

Purpose
It is legitimate to consider introducing a figure whenever you think it could help you summarize or clarify something, or make more efficiently an important point. Figures are ideal for many situations in which words prove inadequate (e.g., characterizing a tissue section)—including the expression of complex ideas, and in circumstances inherently ripe for visual docu-mentation. In scientific text, one very common assignment for figures—specifically, graphs—is revealing or clarifying the nature of a mutual relationship or a tendency.

The "magnetic" effect
Figures invariably act as "eye-catchers", drawing a reader's attention in ways that text cannot. For this reason alone it is important that the pri-mary message you wish to convey with a visual element be made readily apparent. As you work to give substance to a proposed figure, always remember that your goal is maximum effectiveness and ease of interpre-tation.

Figure numbers
In order to facilitate reference to a particular graphic support element from within the text, every figure should be assigned a unique number. As with

tables, each numbered figure must then also be made the subject of at least one explicit text reference. Figure references typically assume a form like

Ex 21–1 … this particular interpretation (see Fig. 3, uppermost curve) …
… have been plotted in Fig. 5–2 with …

In one sense, a figure number occurring in the middle of a text passage could be regarded as a sort of "place-holder" for the figure itself, while serving simultaneously to "anchor" that graphic element in the surrounding "verbal sea". Its function is much like that of the citation markers used to direct a reader's attention to relevant literature references, which are also terse stand-ins: for complete sets of bibliographic information (cf. Unit 15). An even more obvious analogy would be reference (by number) to a specific table (cf. Unit 20).

Few exceptions are permitted to the general rule that every figure must bear an identifying number. A strong case is perhaps most easily made for dispensing with numbers for a very small image (in a mathematical treatise, for example) whose close proximity to a particular text passage is essential, an image that—if surrounded by text—will have little detrimental impact with respect to a harmonious overall page layout. Such an image is described as an "integrated" or "embedded" graphic element.

A figure number is assigned—sequentially, in ascending order—at the first mention of a given figure in the text. Every effort should then be made to position the figure as close to its initial citation as practical.

With a lengthy document like a thesis we recommend that figures receive "double numbers"; e.g.,

Ex 21–2 **Fig. 3–12**. Flowchart for the analytical procedure proposed here.

A composite number of this kind first reveals the identity of the current major subdivision in terms of the overall work (usually via a chapter number) and then expresses the ordered position of that particular figure within the subdivision.

A handy fringe benefit of the double-number approach is that it reduces the likelihood of figure numbers becoming inconveniently large. Confining numerical sequences within specific subdivisions—as suggested—also relieves some of the burden otherwise associated with renumbering in the event that, at a late stage in manuscript development, text is extensively reorganized, or figures are added or deleted. It is *not* a good idea, by the way, to pursue the complex-number strategy to the point, say, of (cumbersome!) *triple* numbers (e.g., "4.3–1") designed to take into account subdivision at a deeper level. Note that if a thesis is so structured that it contains several "Parts" (I, II, III, …), but with chapter numbering continuous throughout, double-numbering of figures will *not* directly reflect "subdivision" at this very highest level.

As implied above, sequential numbers of the type we advocate are re-initialized to "1" with the beginning of each new subunit (i.e., chapter). In other words, the first few figures in the second chapter of a document would carry the designations

Ex 21–3   Fig. 2–1, Fig. 2–2, Fig. 2–3, …

Dashes as separators   The "separator" employed in double numbers is a *dash* rather than a period (or "decimal point") to avoid confusion with section numbers (cf. Unit 8).

Figure captions   A figure number is not considered part of the figure itself, but belongs rather to the figure's *caption*, which is intended to supply the reader with a pithy, verbal characterization of that particular figure's content or purpose. (Some use a different word—"legend"—as though it were synonymous with "caption", but technically speaking the former refers instead to an *elaboration* provided *within* a caption, to be preceded by the figure title; see below).

Captions are generally set directly beneath the corresponding figures, although sometimes one sees a caption *beside* a very narrow figure, aligned with its lower edge (cf. the example near the bottom of Fig. 21–1, for example).

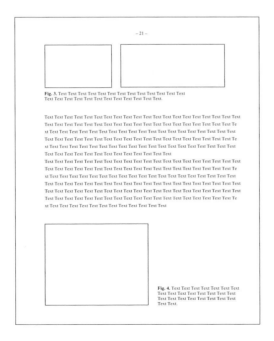

Fig. 21–1. Typical layout of a page containing figures and figure captions.

A proper caption begins with an *identifier*: i.e., the number assigned to that figure, usually expressed in the abbreviated form

Ex 21–4   **Fig. 4–1.**   (occasionally expanded to **Figure 4–1.**)

Some authors substitute a colon for the period that most often marks the end of this identifying element. In-text references to figures also generally reflect the abbreviated notation:

Ex 21–5   … (cf. Fig. 3–1) …   [rather than … (cf. Figure 3–1) …]

Figure title   The most informative part of a figure caption is its summary of the figure's content, embodied in the figure *title*.

As one would expect from its name, a figure "title" is analogous to the title of a chapter, or of a report (or, for that matter, of a table; cf. Unit 20). Reasonable titles for figures might include

Ex 21–6   Schematic diagram of a recycle reactor.
Casein solubility as a function of pH.
Photomicrograph of a yeast cell undergoing division (20× magnification).
Typical distribution of aromatic products derived from X.
Perspective drawing of the zeolite XY framework.

Note that three of these samples include information regarding the technical nature of the parent figure. Of primary concern, however, is an accurate and efficient description of what the figure is intended to show. In common with titles of other types, figure titles rarely consist of complete sentences. In other words, an observation like

Ex 21–7   The collimating lens shown to the right of the slit focuses the light source onto the diaphragm.

would be inappropriate as a figure title.

Figures conveying "primary" information can often be recognized by the presence in their titles of phrases like

Ex 21–8   Electrophoretogram of …
Schematic illustration of …
Radial section of …
Photomicrograph showing …

"Self-evidence"   As previously noted, figures must be carefully conceived and captioned so that they can be readily understood by a typical reader, without consultation of the host document. This obviously means strictly avoiding off-hand comments like

Ex 21–9   For more information, see text.

Legends   Often even a very good title, by itself, provides too little information for full appreciation of a figure's content, as for example in the case of a graph employing symbols that are meaningless in the absence of interpretation.

This is a common problem, and one that clearly requires a solution. Moreover, you might appreciate an opportunity occasionally to expand somewhat upon a particular title so as to direct the reader's attention to a specific detail within the figure, or inject a few key experimental parameters, or identify prerequisites to securing images of the type shown. Situations of this sort are in fact easily dealt with: in a *legend*, appended directly to the figure's title, immediately after the closing period, but isolated from it by a dash. Alternatively, the legend can be formulated as a separate text block set beneath the title. Ex 20–10 provides samples of both types of legend.

Ex 20–10a *Fig. 1–12.* Schematic rendering of a conventional atomic absorption spectrometer.—L Hollow-cathode lamp, S sample reservoir, W wavelength selector, D detector.

b **Fig. 8.** Alkane separation by programmed-temperature gas chromatography.

Column:	50 m methylpolysiloxane
Temperature:	28...200 °C; 2 °C/min
Duration:	46 min
Carrier gas:	hydrogen
Detection:	FID

It is impossible to declare one of these formats to be "best". The block arrangement in the second illustration above is certainly clearer from the standpoint of its information content, but at the expense of greater space demands. A block approach is especially well adapted to a legend that applies to several parts of a composite figure (such as one consisting of several chromatograms), where the reader is invited to make visual comparisons.

Legends can be useful in quite another way as well: they permit one to circumvent extensive labeling *within* a figure. Graphic material profits greatly from the absence (insofar as possible) of embedded text, especially lengthy intrusions by this inherently foreign element. Besides, it is easier to create a pair of legend entries like

Ex 21–11 B bridge resistor
1 reactor

and then supplement the corresponding illustration with the requisite (roman) letter "B" and italic (cf. Fig. 21–6a) numeral "*1*", as opposed to trying to find an aesthetically acceptable way of accommodating entire words in a schematic diagram, especially given the limited amount of free space present in a well-designed figure.

As Ex 21–10b suggests, legends can also be commandeered to divulge key experimental details, but you may find it hard to decide what you wish to communicate in this way, and what to reserve for initial disclosure in the Experimental section of your thesis. Remember that the main purpose of a figure caption (including any legend it may contain) is to make the figure intelligible in the absence of the host document. Try not to go beyond this mandate, concentrating instead on brevity.

If you can, design all your figures such that their legends require only standard alphanumeric characters, as in

Ex 21–12   Experiment 1, Experiment 2, … (or Exp.1, Exp. 2 or simply *1*, *2*)

A few conventional symbols like ●, ▲, ■, ○, △, and □ are also acceptable, however, preferably ones comparable to those shown here (see also Sec. 21.3 below).

Letters (a, b, c, …) are usually employed to distinguish the several parts of a composite figure, which are then further characterized in the context of a single inclusive caption. The latter typically assumes one of the following forms:

Ex 21–13a   **Fig. Z.** Spatial course of XY, expressed a) as a set of recorded data-points, and b) in the form of a smoothed curve.

b   **Fig. Z.** Spatial course of XY. — a) Recorded data-points; b) linear extrapolation (with smoothing).

c   **Fig. Z.** Spatial course of XY.
a) Recorded data-points
b) Linear extrapolation (with smoothing)

(See also Figures 21–4 and 21–6.)

Captions should be readily distinguishable from surrounding body text. This is usually accomplished in published works by setting the captions in significantly smaller type—e.g., 10-pt captions in conjunction with 12-pt body text.

A figure and its caption are treated as a single unit within the framework of an overall document. The combination is traditionally isolated from copy above and below by the equivalent of at least two text lines.

Placement considerations   Try to maintain strict consistency in the way you arrange figures and their captions on your pages. For example, you might opt for uniform left justification, or carefully centering each figure and its caption between the left and right body-text margins.

Attribution   Now for a very important point you may not yet have taken into consideration: whenever appropriate, you must conscientiously incorporate proper figure *attributions*. "Intellectual property" rightfully belonging to others must be scrupulously respected—which here means acknowledged. Thus, in terms of graphic material, if you wish to avail yourself of a pertinent figure you ran across in an external source, copyright law requires you to provide explicit acknowledgement of its origin in whatever caption you devise for it. Source identification for graphic materials should in fact appear not only in figure captions, but also in conjunction with text passages discussing these figures. The following are illustrative of acknowledgements suitable for use in captions:

Ex 21–14 **Fig. 4**. Genealogy of the yeast employed (from Miller [12]).

Fig. 5. Overview of a fluidized-bed combustion facility (site photograph furnished by the ABC Society, Chicago).

Furthermore, you must never put on display material previously published by others until you have actually received written permission to do so—from the publishing house responsible for the original publication, for example. You are expected also specifically to affirm in the text that you have been granted such permission, and by whom.

Graphic material from an outside source that has not been copied directly, but instead modified in some way to make it more suitable for your own purposes, is also subject to acknowledgement; e.g.,

Ex 21–15 … (based on Smith et al. 2002).

An academic thesis or dissertation is of course a very personal entity, so use of technical illustrations derived from outside sources should probably be the exception rather than the rule.

21.2 Line Drawings

From the standpoint of printing technology, a fundamental distinction must be made between two very different forms of illustration: line drawings and halftones. The latter are most often discussed in the context of photographs, but doing so actually obscures the primary difference. After all, a line drawing can be photographed, too, but the result will still be a line drawing. The real difference lies in the fact that a line drawing consists exclusively of "blacks and whites" ("color" and "lack of color"), irrespective of whether the thing depicted is made up of lines or extended defined regions, whereas a halftone permits simulation of gradual transitions between black and white: in other words, gray shades. Halftone technology also underlies reproduction of full-color images, in which colors of every conceivable hue are combined and mixed.

In 90% or more of the situations calling for illustration in documents of a scientific or engineering nature, line drawings are perfectly capable of conveying the essential messages. The stark, abstract nature of a schematic-type illustration actually lends itself extremely well to black-and-white rendition. Only with the more descriptive sciences—biology and geology, for example—does it frequently become awkward or impractical to refrain from use of halftones, in most cases photographs. Even in these fields, however, schematic drawings sometimes can be more instructive than actual photographs (as in a descriptive physiological interpretation of an organ, for example, or portrayal of the fold pattern defining a mountain range), in that they permit one to point the reader's attention more directly toward the features of greatest interest. Support for this contention is furnished by the illustrative treatments featured in countless highly regarded textbooks and handbooks.

Line drawings often are chosen for illustrating scientific material for another reason as well: their greater flexibility with respect to adaptation and modification—and thus their potential for convenient later reuse under different circumstances. Moreover, in contrast to the situation with photographs, problems seldom arise during attempts to duplicate line drawings, whether through photocopying, photographic methods, or offset print. Finally, line drawings are more amenable to the software manipulation that permits eventual incorporation into computer-based text files.

Replication; color   The now ubiquitous copying technique of "xerography" also is an example of black-and-white technology, which means there are few constraints on processing line drawings with ordinary office photocopiers and computer-driven laser printers. In recent years there have of course been dramatic advances in tools for reproducing color copy as well. Photocopiers with outstanding color fidelity have become a commonplace in commercial copy shops everywhere, and color desktop printers trade almost for pocket-change in office-supply stores and through countless online vendors. Developments like these are certain to have a profound influence over time on the nature of theses submitted to universities around the world, although the subject is not one we propose to pursue. We raise it in passing only to suggest you at least take the possible use of color into consideration as you plan your work—obviously after careful review of relevant institutional policies to which you may be subject.

It is today exceedingly rare to find line art prepared manually by the classical methods based on pen-and-ink rendition, sophisticated templates, and high-quality vellum drawing surfaces, a development almost no one would have foreseen thirty years ago. This challenging (and rewarding!) art form thus becomes another topic we have elected to ignore on the assumption that virtually every contemporary reader will turn automatically to computer-based graphics software whenever a need arises to generate copy for a figure. By the way: we encourage you if possible to employ software that specifically supports "PostScript" output because of potential advantages when the time comes to embed your graphics into word-processing or page-layout files.

Be sure you don't allow yourself to be seduced into overexploiting the limitless typographic capabilities graphics software will place at your disposal. In particular, guard against loading your figures with unnecessary "optical ballast". The presence of more than three type fonts, for example, is almost always detrimental to a figure's overall appearance:

Ex 21–16

~~Electrode kinetics~~
HELMHOLTZ DOUBLE LAYER
Polarization voltages
Activation energy
Exchange-current density

Electrode kinetics
- Helmholtz double layer
- Polarization voltages
- Activation energy
- Exchange-current density

Typographic treatments should also be kept generally consistent from figure to figure.

Lettering size   Alphanumeric characters in a printed figure should be roughly comparable in size to text in the associated captions (i.e., with a character height of ca. 2 mm), a point to bear particularly in mind when preparing drawings that will be subjected to photoreduction.

Pencil sketches   Before you actually begin creating your figures (presumably through the intermediacy of a computer-driven display screen), take the necessary time first to assemble a set of reasonably comprehensive preliminary pencil sketches. Such conceptual previews are invaluable for addressing questions like

- Given the approach you have in mind, will the resulting figure be as informative as it could be?
- Have you found the most convincing and efficient ways to communicate what needs to be said?
- Will individual elements contributing to your figures be optimally distributed over the space available?
- Are all your prospective figures capable of assuming relative dimensions that are appropriate, both from an aesthetic standpoint and with respect to the messages you wish to convey?

– Might some of the figures you envision turn out to be too wide—or too narrow—in terms of overall page design?
– Could any of your planned layouts strike the reader as excessively "busy"?
– Will it be possible to draw everything to an appropriate scale?
– What sorts of labels will be required?

Issues of this sort deserve serious attention before you ever commence to "draft" your figures electronically. Obviously the computer screen itself could serve you as a sketchpad in this context, but most people experience a greater sense of freedom, and less inhibitions, when working with pencil and paper. In the long run, however, all that really matters is achieving final results you consider to be "right" in every sense.

21.3 Diagrams and Graphs

Diagrams in general Probably more than half the figures appearing in scientific and technical publications are dedicated to depicting functional relationships, where values associated with one physical quantity (experimental measurements, for example) are plotted against corresponding values for another quantity. Many different models have been developed for presenting this type of information visually. Individual data points, for instance, might serve as the basis for a scattergram, but such points can also be connected to suggest curves; or one could plot the information as a (presumed) continuous function. In rare cases it makes most sense for data points to be joined in such a way as to produce polygons.

Data points Many types of symbols can be called upon to represent data points in a graph; e.g.,

Ex 21–17a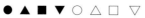

The ones shown here have the advantage that all have traditionally been readily available to the average typesetter (as part of collections of what are curiously called "dingbats"). It is recommended, by the way, that symbols for this purpose be made comparable in size to a text letter "o".

Standard deviations associated with data points are conveniently indicated by slight modification of the same symbols:

b

Line graphs
A line graph is often an exceptionally good medium for addressing questions (and answers) of the type scientists regularly encounter. In a biomedical context, for example, values of a dependent variable (y) displayed in such a graph might reveal the effect of changes due to some independent variable (x) of immediate interest—a significant therapeutic treatment, perhaps, or a phenomenon, or a circumstance. In other words, the diagram would be a direct response to the question—"What happens to y when x changes?" A title for the figure might be of the type

Ex 21–18
The influence of x on y in Z.
Dependence upon ... of ... at ...

Coordinate systems
This type of information is commonly set forth in a framework of Cartesian coordinates,[*] with two (usually perpendicular) axes, a coordinate grid, and an overlay of one or more data curves, all appropriately annotated. The various lines required should ideally have intensities that reflect their relative importance, most easily achieved by varying the line width. Those lines most crucial to the reader, the data curves, should also be most prominent, followed by ones representing the coordinate axes. Tick marks along the axes and other subordinate markings play only a supportive role, and should be correspondingly less pronounced.

Line widths
Many recommend that these three basic line types vary in width approximately in the ratio

$$\text{coordinates} : \text{axes} : \text{curves} = 1 : 1.4 : 2$$

Families of curves
Sometimes a graph will need to feature not just a single curve, but several: a *family* of curves. Each of the lines must then be clearly labeled in terms of the distinguishing parameter, typically through an alphanumeric tag. *Numbers* for this purpose should be italicized (to prevent confusion with numerical information supplied in conjunction with axis scales, for example), but letters should be made upright (roman)—again to prevent confusion, this time with (italicized) mathematical variables. Tags would be explained in a legend included as part of the figure's caption.

In many graphs one can fairly easily plot several different dependent variables in an unambiguous way, all with reference to a single independent variable, but feel free if necessary to call upon an assortment of line types to help assure clarity. Recommended alternatives to the usual continuous

---

[*] The term "Cartesian" is derived from the name of the French philosopher René DESCARTES, who is usually credited with the notion of data depiction in terms of two perpendicular axes, although it is not actually essential that these be perpendicular. Consider, for example, the longitude and latitude lines so central to cartography (e.g., in geographical and meteorological maps).

line include dotted or dashed lines as well as hybrids of the "dot-dash-dot" variety:

Ex 21–19   ——————   — — — — —   ············   —·—·—·—··

Regions In rare cases, significance will be associated not only with data curves, but also with regions they appear to define. Extended areas within a graph can be highlighted or differentiated with the aid of *patterns*: cross-hatching, screens, or other distinctive backgrounds (which might even involve selective use of color).

Arrows, etc. Several possibilities exist for indicating the meaning of specific locations or regions in a diagram or graph. The most common rely on lines, arrows, or selective screening (cf. Fig. 21–2).

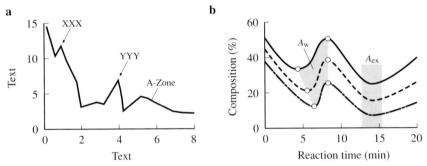

Fig. 21–2. Calling the reader's attention to special features or regions in a graph. – **a** By use of reference lines or arrows; **b** with screening.

Axes Numeric values associated with the independent variable in a typical graph increase from left to right along the horizontal (or "*x*") axis, the *abscissa*, whereas those for the dependent variable increase upward along the vertical ("*y*") axis, the *ordinate*. A direction of increase can be indicated explicitly with an arrow incorporated into the axis label (cf. Fig. 21–3a), or by installing an arrow point at the end of the axis itself (Fig. 21–3b). Such devices are of course redundant in the presence of numerical scale markings, where their inclusion is actually discouraged. On the other hand, directional arrows are indispensable with strictly qualitative representations like those in Fig. 21–3.

Tick marks Axis scales are calibrated with the aid of "tick marks" to indicate specific numeric values. The marks themselves might be regarded as vestiges of a (now implicit) coordinate grid. Generally such marks are directed toward the interior of the graph (i.e., to the right from the ordinate, upward from the abscissa). In the event that this would prove disruptive due to a particular graph's content, however, outwardly-directed marks are also per-

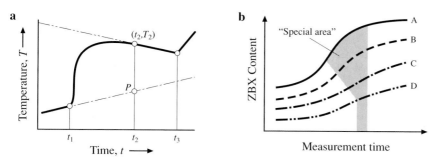

Fig. 21–3. Qualitative depictions. – **a** Arrows embedded in the axis labels; **b** axes terminating in arrow points.

missible. Be careful that your tick marks are not set too close together, and that all marks are of the same length (with the possible exception of extra long markings to indicate important regular intervals). Representative examples are provided in Figure 21–4. It is not essential, incidentally, that all tick marks be elaborated with explicit numerical values, but at least the first and last should be.

Coordinate grids

Standard axis markings are occasionally extended in such a way that they essentially reestablish a tangible coordinate grid (cf. Fig. 21–5), although the practice is generally frowned upon. "Framing" of a graph through the addition of right and top boundaries is similarly discouraged. The problem in both cases is that the presence of extra lines, which are essentially superfluous, can have the effect of distracting a reader from the actual message a graph is supposed to convey. Only when it is important that one be able to estimate plotted values fairly accurately at a glance should extra horizontal and/or vertical guidelines be provided.

Zero points

Always label the zero points of your coordinate systems with explicit zeroes along both the ordinate and the abscissa, even when the two coincide

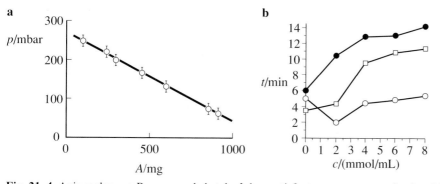

Fig. 21–4. Axis scales. – **a** Recommended style; **b** less satisfactory arrangement (in that tick marks are directed outward, inconsistent in length, and too close together).

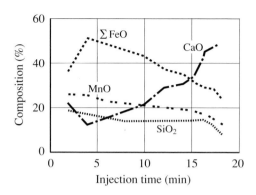

Fig. 21–5. Axis scales elaborated to produce an overt coordinate grid.

(cf. Fig. 21–6a). While the region in the vicinity of the value "zero" may be of no particular interest to you in the case of at least one of the axes, it should still be shown and marked so as to completely rule out misunderstanding. The corresponding axis can then be interrupted as a way of producing a more convenient picture (see Fig. 21–6b). Alternatively, separate the scale information for the two axes completely, as has been done in Fig. 21–6c.

Negative values   If an axis happens to extend into the region of negative values, *all* affected markings should be labeled with preceding minus signs.

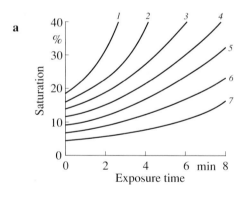

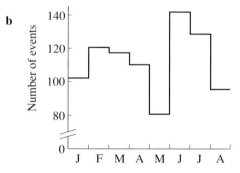

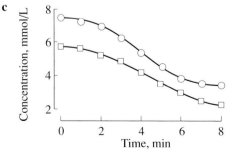

Fig. 21–6. Clear designation of the zero points associated with pairs of coordinate axes. – **a** Two zero points that coincide. **b** interrupted ordinate; **c** completely independent scales.

Typography
Numerical axis calibrations should be shown in ordinary (roman) type, and all numbers should be oriented on the page in the same way as body text. In other words, value indicators applicable to the ordinate must not be rotated to conform with the axis direction. A reader should thus *not* find it necessary to reorient the page in order to see the numbers clearly (cf. Fig. 21–7).

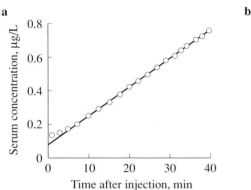
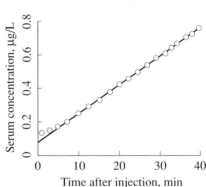

Fig. 21–7. Axis-scale calibrations. – **a** Correct; **b** incorrect.

Units; axis notation
In a scientific context, the specific numerical values associated with a quantity almost always depend upon the unit selected, so it is essential that units be identified in all graphs, just as they must be in tables. One common approach to introducing unit symbols is incorporating them into the axis-scale information. Annotations for this purpose should be placed toward the right-hand end of the horizontal axis and near the top of the vertical axis, in both cases between the last two markings. If space proves to be a problem, the next-to-last (or even third-from-last) number can simply be omitted (cf. Fig. 21–8)

Under no circumstances should a quantity symbol be accompanied by a unit symbol in parentheses! Parenthetical notation is perfectly acceptable, however, when it serves as an elaboration on the *name* of a quantity; e.g.,

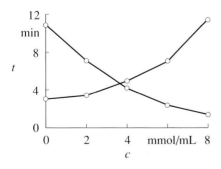

Fig. 21–8. Display of unit symbols in conjunction with a graph.

Ex 21–20 Luminance (cd m^{-2})

A unit symbol following a quantity name may also be separated from the latter by a *comma*, as in Fig. 21–7.

Two other possibilities should be mentioned as well for identifying units of measurement: adding the word "in" after a quantity symbol, followed immediately by the appropriate unit symbol (e.g., "U in V"), or declaring the quantity and its units in fractional form (e.g., U/V); in this case the context (cf. Unit 20) is no longer that of values of a quantity, but rather pure numerical values—just as these are indicated on the axis scale, a form of notation illustrated in Ex 20–9 and Fig. 21–4.

Try to avoid the necessity of including power-of-ten factors in your axis labels by prudent selection of unit prefixes. One permissible exception relates to an axis marking like

Ex 21–21 10^6 g^{-1} s^{-1}

in which case the quantity of interest will have been plotted in terms of the "unit" 10^6 (radioactive decay events, for example) per gram and second.

%, ‰, ppm,
ppb, ppt
A treatment analogous to that proposed for the order-of-magnitude factor in the preceding example also applies to symbols for percent (%), "per mill" (‰), and relationships of the parts-per-million (ppm) type, all of which also would be announced between convenient pairs of scale numbers.

°, ′, ″
On the other hand, the angular magnitude symbols associated with the degree (°), minute (′), and second (″) are treated in the rather unusual way of appending them to every calibration value indicated explicitly along an axis.

Quantities are preferably specified in terms of their symbols rather than by name. Be sure in all your figures, just as in text, that quantity symbols are set in italics.

If a graph in itself contains no symbol to identify a quantity whose dependency you are trying to illustrate, the name of that quantity—or, if necessary, an equivalent mathematical expression—can be included in the appropriate axis label. Sometimes essential information is best supplied in a more verbose way:

Ex 21–22 Oven temperature in °C
CO_2-consumption per kg of medium in mmol

One seldom encounters serious problems accommodating such explanations along a horizontal axis, but the vertical axis is another matter. If you were to orient text for this purpose like other text on the page, the figure as a whole would quickly become too broad. When a long axis label is absolutely indispensable in this setting, rotate it parallel to the axis, with text running from bottom to top (i.e., so that to see it clearly a reader would need to turn the page 90° to the right: clockwise), as in Fig. 21–7a.

Multiple quantities   Occasionally one would like to plot behavior related to more than one quantity (concentration, electrical conductivity, and light absorption, for example) using a single graph. This of course means you need multiple axis scales, but ambiguity must be rigorously excluded. The most obvious solution in the case of *two* quantities is use of the right (inner) side of the ordinate for one scale, and the left (outer) side for the other; or a second scale can be established along the right-hand side of the graph. Yet another possibility would be to introduce a second auxiliary axis, with a different scale, alongside the principal axis (cf. Fig. 21–9).

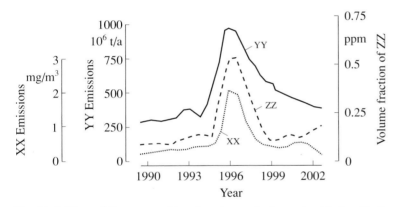

Fig. 21–9. Three different quantities plotted on a single graph with the aid of auxiliary axes.

Spectra, chromato-   Apart from the functional graphs that have so far occupied nearly all our
grams.   attention, other sorts of figures deserve comment as well. For example, it is not uncommon to wish to reproduce in a thesis various spectra, chromatograms, or analogous types of instrumental data output. Such material will generally require a certain amount of processing before it is ready to become part of a formal document. The usual approach is to *scan* one's originals and then edit the resulting scan files using a suitable graphics program, thereby generating images of an appropriate size and quality for subsequent electronic "pasting" into a word-processor (or page-layout) file.

Flow charts   Another useful type of figure the preparation of which requires relatively little creative or artistic talent is the *flow chart*, where some process, or perhaps an organizational scheme, is depicted in a strikingly abstract way (cf. Fig. 21–10). Apart from a bit of imagination, producing such a chart

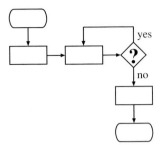

Fig. 21–10. Abbreviated version of a process scheme or decision algorithm in the form of a flow chart.

requires only the drawing and labeling of an appropriate array of horizontal, vertical, and oblique lines, together with a few square, rectangular, circular, and/or elliptical boxes. Software is widely available to relieve you of most of the associated burden.

Other types of figures lend themselves to similar skeletal expression—including graphic representations of engineering facilities, laboratory setups, analytical procedures, and the like. Again, rectangles typically play a central role, as substitutes for specific components or operational steps. These are once more joined in a meaningful way by lines and arrows, and then suitably labeled, preferably taking advantage insofar as possible of concise, conventional symbols.

Bar charts, pie diagrams

Popular graphic devices in conjunction with work of a *statistical* nature include the bar chart and the pie chart, examples of which are shown in Fig. 21–11.

Graphs of this sort assume a more expressive character when specific segments are singled out through distinctive surface patterns, perhaps simulated gray shades. Gray effects are usually achieved with sophisticated arrays of dots, although line patterns sometimes are called upon instead. Most computer software for preparing drawings places at the user's disposal a wide variety of standard patterns, which one can then apply anywhere desired with a mere click of the mouse.*) Much drawing software is incredibly powerful and versatile, but at the same time surprisingly straightforward to use. Picking out a suitable program is of course the first step. Our only recommendation in this regard is that you concentrate on software for producing *vector* graphics (preferably of the PostScript variety) rather than bitmap images, since this will help ensure high-quality reproduction at whatever scale is necessary.

---

* If the samples we show here seem too prosaic, examine the consequences when you instruct your own software to add three-dimensional effects to some of your charts.

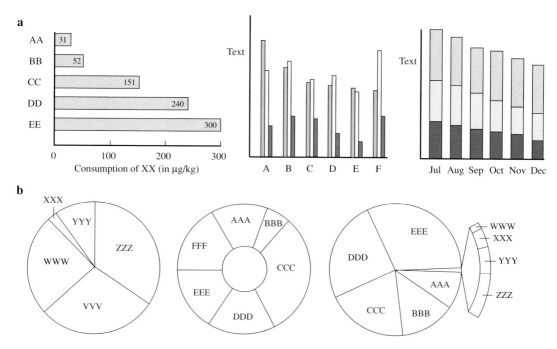

Fig. 21–11. Examples of semischematic and schematic graphs. – **a** Bar charts; **b** pie charts.

21.4 Halftones

Photos   If you decide to include in a thesis visual documentation of your actual research activities you will in all probability face the challenge of working with photographs: photomicrographs derived from tissue samples, perhaps, or habitat overviews. It is essential that the images selected depict the item(s) of interest clearly, vividly, and preferably more or less in the center.

Important details in photographic images are sometimes singled out by enclosing them in prominent circles, or by directing superimposed arrows toward them. Consider also providing the reader with a sraightforward way of acquiring a sense of scale.

Processing   The traditional approach to incorporating materials of this kind into a thesis was always to prepare a set of high-quality, high-contrast glossy originals, which were then simply pasted into spaces reserved for them in the printed document—often in the master copy alone, duplicates being augmented only with lower-quality photocopies. Today it is more common to create electronic versions of photographs—with the help of either a scanner or a digital camera—for including directly in one's word-processor or page-layout data files. This route obviously offers a number of advantages, not the least of which is the opportunity to edit the images (digitally) as extensively as you wish.

Printing technology has in recent years advanced to such a point that there is no longer any reason to be squeamish on this account about illustrating a thesis with halftones—or even full-color images. In all probability doing so will have some impact on printing costs, however, especially since you will probably wish then to take advantage of special paper—although this could depend on official standards established by your university.

C 21–1   What is involved in "anchoring" a figure in the text?

C 21–2   How does a "legend" for a figure differ from a "caption"?

C 21–3   Describe how a figure caption can be used to simplify the labeling of a figure.

C 21–4   What is the fundamental difference between a line drawing and a halftone?

C 21–5   Cite a common rule of thumb for relative line widths in a graph.

C 21–6   Suggest type-style and dimensional guidelines applicable to numbers, letters, quantity symbols, lines, and arrows included in figures.

C 21–7   Explain the various techniques for introducing unit symbols into a graph.

C 21–8   How might each of the graphs shown below be improved?

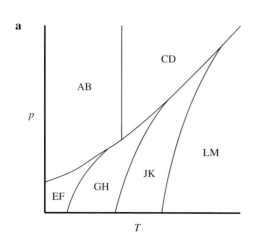

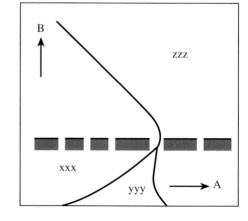

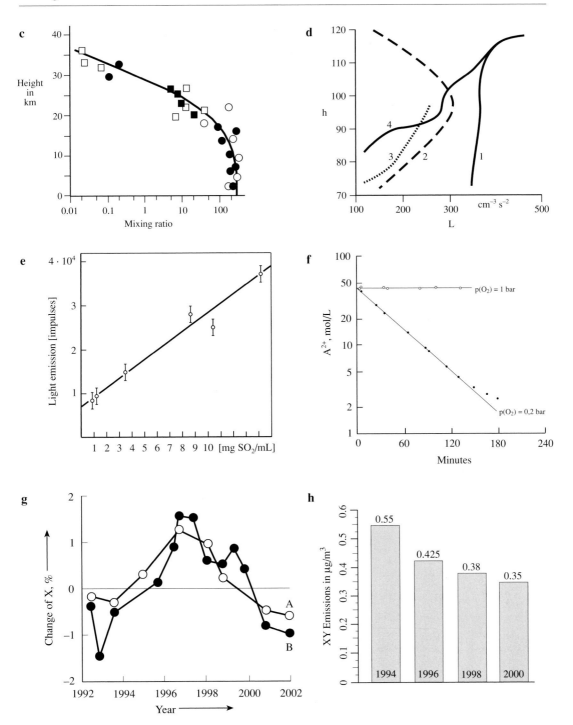

Solutions
to the Challenges

Note

We have tried to keep our "Solutions" to the "Challenges" as concise as possible, and strongly recommend that you not only compare your solutions with those proposed here but also closely examine once again the original challenges themselves. In some cases, we have refrained from providing a solution: typically with challenges related directly to the reader's own research experiences, or where providing a response would entail simply reproducing a passage from the text (although some repetition of this sort has nonetheless survived). Marginal pointers in the text should facilitate locating important information relevant to specific challenges.

Unit 1 Solutions

S 1–1   It is a convenient way of defining and labeling individual experiments (consistent with a notebook's organizational role!), and it facilitates internal cross-referencing as well as the labeling of materials related to the experiments.

S 1–2   Restrict yourself to permanently *bound* notebooks; number all pages; ensure that entries are continuous (no blank space!) and that each begins on a new page; leave no pages blank, and "cross out" any unused portions of a page; enter everything with a ballpoint pen (no pencil entries!).

S 1–3   In order to rule out the possibility that information was entered after the fact, thereby enhancing the "documentary character" of notebook entries.

S 1–4   This makes it possible to reconstruct the workflow sequence, which may be important in establishing claims of priority. In a sense it also allows the notebook to serve as a sort of "time sheet" or work record.

S 1–5   The fact that pages can easily be added to or removed from a ring binder voids all assurance of the integrity of the written record.

S 1–6   Certainly not. You must yourself define the starting and end points of each "experiment", relying strictly on pragmatic criteria.

S 1–7   Two (facing) pages.

S 1–8   Completeness and clarity are of primary concern, along with legibility, all from the point of view of someone *else* who is interested in examining the record. Within these constraints, brevity is also desirable (e.g., there is no need to express yourself in complete, grammatically sound sentences).

S 1–9   "Jargon" is perfectly acceptable so long as it will be widely understood. (Examples will be discipline-specific.)

S 1–10   To facilitate locating a particular piece of work (e.g., through a table of contents).

S 1–11   Assured authenticity and timely origin, comprehensiveness, and reproducibility of the results.

S 1–12   First part: A title, date/time and (if appropriate or not self-evident) place information, a rationale for conducting the experiment, and notes regarding relevant circumstances (e.g., necessary materials and supplies, special equipment, etc). Second part: description of what you actually did and what in fact transpired. Third part: results.

S 1–13   (Individualized, discipline-specific response.)

S 1–14 In search of experimental details, or to repeat and/or verify reported findings.

S 1–15 The report shows considerable linguistic "polish" (e.g., everything is now expressed in complete sentences), certain pieces of raw data have been translated into more meaningful form (e.g., a crude weighing into a percent yield), and some minor details have been omitted.

Unit 2 Solutions

S 2–1 A collection of bibliographic information; data characterizing individual elements within an assemblage of references, possibly together with content summaries or even full-text copies. Reference file, literature support, documentation.

S 2–2 Classification by subject requires subjective decisions, such as choices among terminological alternatives, whereas author names are unambiguous.

S 2–3 For cross-referencing within a literature file, linking entries in a bibliographic data file with actual documents (reprints, excerpts, personal summaries), functioning as "place-holders" in early drafts of documents.

S 2–4 Essential: complete bibliographic information (sufficient for retrieving the source from a library collection, for example); optional: keywords, abstracts.

S 2–5 Computer-based data management, with support from a database program offering flexible search options.

S 2–6 (Individualized, discipline-specific response.)

Unit 3 Solutions

S 3–1 Laboratory notebooks and associated records (e.g., spectra), interim reports, complete bibliographic information and relevant literature summaries and/or reprints. Above all, prepare backup copies on a regularly basis for storage in a separate, safe location.

S 3–2 Writing should be deferred until the active research phase is completely behind you (not only in your opinion, but from your adviser's perspective as well!). Formal writing should be preceded by development of a comprehensive outline. University calendars and an advisor's travel plans are among the many external factors that can affect scheduling.

S 3–3   A solution to this challenge is presented in the text of Unit 8 in the form of Ex (Example) 8–7.

Unit 4 Solutions

S 4–1   (Individualized, discipline-specific response.)

S 4–2   Investigated or studied; frozen or crystallized; improved or recovered.

S 4–3   Should evaporate it as rapidly as possible; began determining X with Y; A preferentially couples in this case with B; expansion of U with V was seriously considered; began to discuss the general suitability of such a procedure; analyzed Z spectroscopically.

S 4–4   Because of the change in x activity at pH > 8 we decided to ...; or: the activity of x falls off dramatically at pH > 8, so we ...

S 4–5   "In good yield" is sufficient, and avoids the more judgemental adjectival construction; more professional would be something like "in high yield (88%)".

S 4–6   The nutrient content of barley differs from that of corn.

S 4–7   The corrosion resistance of XY was studied in a thermostated experimental chamber. Structural components were found to corrode in moist air only in the added presence of sulfur dioxide or nitrogen oxides. This research was conducted as a part of the Defense Department research initiative "Long-Term Storage Stability of XY Structural Components".

S 4–8   ... It is now commonly assumed that preventing contamination with other solid and solvent wastes facilitates recovery and recycling of chlorinated hydrocarbons to a much greater extent than was originally imagined.

... In the meantime, distributors and consumers have become aware of new trends with respect to both computer hardware and software. Unfortunately, contrary to expectations, life has not become correspondingly simpler for the user; indeed, users must become more engaged than ever if they wish to understand what is possible and how best to proceed.

... When preparing this amino acid it is important to pay close attention to both pH and the rate of (dropwise) diamine addition, since otherwise the yield of the desired product falls dramatically (due to a side reaction; see Equation 12).

S 4–9   There was then added 10% serum from a newborn calf and 1% essential amino acids.

(This makes clearer the meaning of the percentages.)

S 4–10   ... the method, discovered four years ago ...
... the avian flu virus, which had proliferated inconspicuously ...
... the rising temperature ...

Unit 5 Solutions

S 5–1 Assemble at the outset all the necessary background materials, and then concentrate on establishing—in outline form—a smooth, logical flow of ideas.

S 5–2 A rough draft should be prepared rapidly, so one must anticipate that it will not read very well even though the underlying structure should be sound. With each revision, expression of the author's ideas should become more focused and precise, and the final wording should gradually begin to take shape. Nevertheless, every reading will reveal new rough spots to attack. The "final draft" (perhaps the sixth or seventh!) will be the first one you find you can read (thoughtfully!) without feeling a need to make further changes. Transforming this into "final copy" entails resolving all outstanding formatting details and generating what you can conscientiously describe as "the perfect thesis".

S 5–3 Exceptionally easy revision of every aspect of a document, rapid "navigation" capability, means for instantly finding any specified word or phrase, reliable spell-checking, facile preparation of backups, enormous flexibility with respect to type styles and sizes.

S 5–4 By use of different type styles (boldface and italic variations) and/or different type sizes.

S 5–5 One-and-one-half to two times as much space before as after.

S 5–6 We would cite the following as perhaps some of the most important advantages:

1. Digital storage is far more space efficient.

2. An "electronic thesis" can be accessed without delay, and distributed both more rapidly and more widely than one could ever have foreseen in the past.

3. A digital manuscript can easily be enhanced and enriched with digitized auxiliary materials (e.g., images, both static and animated; sound).

4. Auxiliary materials like those mentioned can actually be *integrated* into the thesis.

5. Preparation of such a document provides valuable training for likely future responsibilities and opportunities.

6. The likelihood that theses will become more widely appreciated as valuable sources of information.

7. Availability of tools for creating dynamic links ("hyperlinks") both internal to a thesis and to the vast wealth of resources residing on the Internet.

8. The potential for carrying out complex and instantaneous searches that even take tables and figures into account.

Unit 6 Solutions

S 6–1   Essential elements: complete title and nature of the document, author's name, submission date; frequently included: identity of the host institution and/or department, copyright notice.

S 6–2   If the title would otherwise be too long (more than about 10 words).

S 6–3   **a** through **f**: all phrases like "Investigation of …", "Report on …", etc., are superfluous; **b**, **d**, **e**, and **g**: a subtitle would in each case make it possible to move important words toward the front.

S 6–4   See Fig. S–1.

Unit 7 Solutions

S 7–1   The best place for a dedication is usually the first right-hand page after the title page, perhaps to be followed on the next right-hand page by acknowledgements and/or a preface.

S 7–2   They should be included in the normal pagination of the document, but should display no explicit page number.

S 7–3   This sample preface is insufficiently structured, a problem that is easily resolved, however; e.g.:

Preface

The work described here was carried out in the Stearns research group of the Department of Biology at XYZ University between July 2001 and November 2003.

I wish to thank Prof. J. B. Stearns for suggesting that I undertake an independent investigation in this area, and for all the support he provided.

George Walters, together with the other members of the group, contributed regularly to my progress through countless informal discussions. David Weizman in particular supplied valuable assistance in the recording and interpretation of esr spectra. I am delighted to have this opportunity to express my sincere appreciation to them all.

Finally, I am grateful for generous fellowship support from the Akron Institute during the period July 2001–November 2003.

Westville, in November 2003   Arnold Abrams

Calculation of Vibrational Spectra for
Crystalline Hexamethylbenzene
and Perfluorohexamethylbenzene

Masters Thesis

submitted by

Herman L. Johnson

2003

**Effect of Vinpocetin on the
In-Vivo Flexibility of Erthrocytes**

Measurement Based on a New Centrifugation
Technique

by

David L. Jones
Boston, Massachusetts

2004

Thesis Submitted for a Master of Scuence Degree

Fig. S–1. Revised title pages (see Challenge C 6–4 for the Originals).

Unit 8 Solutions

S 8–1   Sec. 3.1.1 should not be deferred to p. 38. Similarly, there should be no gap preceding Sec. 3.2.1. Presentation of page numbers should be strictly uniform (i.e., that for Sec. 3 should appear at the right margin, introduced as usual by a dotted line), and all dotted lines should take the same form (preferably analogous to the line accompanying the entry for Sec. 3.1.3, for example). Finally, the "Introduction" after Sec. 3.2 must be incorporated into the numbering scheme.

S 8–2a   The differences should be clearly apparent from the following suggested alternatives:

1	Introduction
2	Results and Discussion
2.1	Reactions in Polar Organic Solvents
2.1.1	Metal–Hydrogen Exchange
2.1.2	Metal–Halogen Exchange
2.2	Reactions in Non-polar Media
2.2.1	Condensed-Phase Reactions
2.2.1.1	Photolysis Experiments with Cyclohexane as Solvent
2.2.1.2	Photolysis Experiments with Benzene as Solvent
2.2.2	Reactions in the Gas Phase
3	Experimental

etc.

S 8–2b   Another possibility would be to use roman numerals to divide the thesis into "parts" (cf. Ex 10–5):

I Introduction
II Results and Discussion
1 Reactions in Polar Organic Solvents
1.1 Metal–Hydrogen Exchange
1.2 Metal–Halogen Exchange
2 Reactions in Non-polar Media
2.1 Condensed-Phase Reactions
2.1.1 Photolysis Experiments with Cyclohexane as Solvent
2.1.2 Photolysis Experiments with Benzene as Solvent
2.2 Reactions in the Gas Phase
III Experimental
3 Metallating Agent

...

4 Photolysis Experiments
4.1 Apparatus
4.2 Workup

[The second example has been expanded somewhat relative to the original version to illustrate more clearly the notion of continuous chapter numbering in the context of "parts", and to indicate also how the Experimental section might be subdivided.]

S 8–3   It looks perfectly reasonable—for a thesis in a field like physics, for example.

S 8–4   Chapter 2, Theory.

S 8–5   Numbering in this sense serves to make the hierarchical structure of a document more obvious. Furthermore, it greatly facilitates cross-referencing within the text (e.g., "see Sec. 5.2").

Unit 9 Solutions

S 9–1   The first paragraph should have been made part of a preface. Paragraph two, on the other hand, can be allowed to stand as it is. A possible reformulation of paragraph three might read:

Product formation occurs according to the scheme

$$XY–R + Z \longrightarrow Z–R + XY$$

providing a much simpler pathway to compounds of the type Z–R than any previously reported.

(Note the elimination here of all direct reference to both tables and the published literature.)

S 9–2   The following abstract proposal contains all the significant information cited:

The objective was to determine whether charcoal in an exhaust-gas filter for removal of ZZZ can be replaced by porous Y-silicate.

If, prior to its utilization, the Y-silicate is tempered for ca. 30 min at 500 °C, 90% of the ZZZ present is absorbed (charcoal: only 75%). Furthermore, unlike charcoal, residual Y-silicate after utilization need not be handled as a "hazardous waste" due to very low solubility of silicate-absorbed ZZZ.

S 9–3a   Vague and ambiguous, in that one never learns precisely what was done!

S 9–3b   Concise and clear, and thus seemingly quite satisfactory.

Unit 10 Solutions

S 10–1   The expression "our laboratory" would be inappropriate in a thesis (a better approach is to cite specific publications attributable to the group leader). Describing at the very outset the results obtained is also problematic, as is the paucity of references to the published literature. Consider the following alternative formulation (which of course still suffers from the last-cited deficiency):

This work was undertaken to establish whether it might in fact be possible—contrary to the observations of Young and Smith [4]—to obtain compound X from the industrial byproduct YY.

S 10–2   The first example (Y) of a compound of the class XXX was prepared in small quantity as early as 1921 (Mayer 1921). Thanks to a new synthetic pathway via Z, Miller and Jorgenson (1955) finally succeeded in obtaining substantial amounts of Y. Chiang (1958) studied IR and UV spectroscopic properties of the compound, as well as its fundamental chemical reactivity. Y subsequently revealed its utility in the pharmaceutical field for preparing compounds of the type DDD (Peters 1961), although commercial exploitation only became truly feasible after conditions were established for reducing (to ca. 10%) the formation of byproduct A (Miller 1969).

   The current project was initiated with the goal of carrying out laboratory-scale studies to identify the best approach toward further decreasing the extent of A in the product, with special emphasis on varying the temperature and solvent.

S 10–3   Generally speaking, this introduction can be described as rather successful. It comes to the point quickly (in ca. 200 words, with 7 references), is well structured (as four short paragraphs), and conveys an impression of "significance". What was and was not previously known has been presented just as clearly as the resulting challenge (How can one best help patients displaying only marginal cardiac performance?).

   On the other hand, at least from the perspective of an "outsider", it is never made really clear what precisely is meant by "cardiac insufficiency"—certainly a shortcoming worth remedying. We also feel that the reader is left too much in the dark about how the associated hematorheological studies were conducted. Finally, we would be inclined to restructure the second sentence in the second paragraph to begin: "While others have reported ...".

Unit 11 Solutions

S 11–1   It facilitates clear separation of (1) what was previously known, (2) your own findings, and (3) commentary on the work.

S 11–2   Certainly the results due to others, and the interpretation of an instrument reading would *not* belong in a Results section, but rather in a Discussion.

S 11–3   (Individualized, discipline-specific response.)

S 11–4   Past tense: describing specific research activities; present tense: reporting of observed or established characteristics, and reference (in sentence form) to sundry overall aspects of your research.

S 11–5 The second and third sentences describe "results", whereas the first consists largely of material better suited to the Experimental section. The last sentence belongs in the Discussion.

S 11–6 The first sentence is dedicated to the most significant finding—clearly enunciated. Less important results were deferred to the second sentence ("most important → less important"). The last sentence includes an assertion whose validity remains in question, so this should probably be moved to the Discussion.

Unit 12 Solutions

S 12–1 The Discussion is the place for you to *comment* on your work: to provide analysis and interpretation, to elaborate on how your findings fit into a broader context, and above all to make a strong case for treating your work as meaningful. On the other hand, the Results section should be a rather stark account of observed phenomena, lacking all taint of after-the-fact reflection.

S 12–2 Compilation, analysis, commentary, explanation, attribution, comparison, evaluation, context-setting, classification, demarcation, and extrapolation.

S 12–3 By rigorous documentation of the source of all ideas and results drawn from others, as well as occasional interjections along the lines "personal experience has shown …".

S 12–4 The present tense is appropriate for something (ongoing) that *is*, just as the past tense applies to something that *was*. Perfect tenses (I *have* seen, I *had* gone) are distinctive for their deliberate vagueness with respect to timing.

Unit 13 Solutions

S 13–1 No, any "conclusions" you wish to draw could instead be made a part of the Discussion.

S 13–2 Immediately following the Discussion.

S 13–3 The focus tends to be on the broader significance and context of one's results, and perhaps logical follow-up measures. (Individualized, discipline specific response to the second part.)

S 13–4 The author's "conclusions" are certainly highlighted. Phrases like "In summary, …" and "may now be taken as established" contribute to making the purpose of the statement apparent. One presumes that the original question underlying the research was something like "Is there a better growth medium for mammalian cells than FCS?" A distinctly positive answer to the question is obviously being reported, although one significant limitation is also noted ("not … with all cell lines"). With "… must be considered independently for each cell variety", incentive is provided for further research. The scientific domain to which this work belongs is also perfectly clear (cell culture). Less straightforward, however, is what the author means by "support growth *much more effectively*"; i.e., what parameter(s) did the study measure? It is also a bit disconcerting that, from this passage in isolation, one comes away with no sense at all of the nature of the "alternative sera" investigated, something that presumably could easily have been communicated. On the other hand, the last sentence is quite good, in that it underscores the significance of the findings. This work can be regarded as an example of a methodological study which has led to a new method more economical than the standard approach. One might argue that the adjective "interesting" interjects a superfluous bit of escalation, but in general the tone of the overall passage and the choice of terminology is laudable; even someone with no background whatsoever in the field could feel enlightened after a single reading. Note in particular that the author was careful to explain the possibly unfamiliar abbreviation "FCS".

Unit 14 Solutions

S 14–1   (Individualized, discipline specific response.)

S 14–2   You must be careful to include everything one would need to know (regarding, for example, materials, apparatus, experimental conditions, etc.) in order to judge and also to duplicate the reported work, along with the actual results that were observed. Every experiment playing a role in the "Results" section must be included, although illustrative examples would suffice in the case of a long series of equivalent procedures. Experimental findings can also be provided in summary form (e.g., in tables). Procedures obtained directly from the published literature need be described only to the extent that changes have been introduced.

S 14–3   "Raw data" refers to information in precisely the form originally obtained, and it is expected that data of this fundamental sort will be heavily represented in the Experimental section of a thesis. "Derived data", on the other hand, is information that has been subjected to some sort of processing, such as being recast in a different form. Derived data may be made part of either the Experimental section or the Results. Care should be taken clearly to explain any conversion factors, corrective terms, or equations you resort to in the course of data transformation.

S 14–4a The first two sentences should certainly be made a part of the Discussion rather than appearing in the Experimental section. A rendition like the following would be more appropriate here:

> In order to conduct a desorption measurement like those described by Crank and Parl,[13] as well as Stuart,[15] a glass vessel containing the sample is suspended from a spring balance, and ensuing loss of weight over time is monitored directly.

S 14–4b The description provided is too vague. What was the concentration of the solution? How was warming accomplished? What filtration technique was employed? How much time was allotted for drying—and drying by what method? A more satisfactory account might read as follows:

> **Preparation of XXX:** A solution of 250 mg of AAA (moisture sensitive!) in 30 mL of BBB is heated on a steam bath (round-bottom flask, protected by a drying tube) to 90 °C. In the course of ca. 1 h a yellow precipitate appears, which is then collected in a fluted filter and left for 2 h in a drying oven at 120 °C (yield: 190 mg, 80%).

[This account of presumably original work is somewhat unusual for a thesis, by the way, in that it has been cast in the present tense, creating an air of certainty more in keeping with a well-established procedure. If for some reason you were also to adopt this mode of expression, be consistent, and strictly avoid jumping back and forth arbitrarily between the present and the past.]

Unit 15 Solutions

S 15–1 – Journal title not abbreviated (1) or abbreviated incorrectly (2);
– academic titles are never included in references (or in running text) (1 and 5);
– first names are not written out in full, as has been done in (2); initials are missing in (1, 4, 7);
–the journal in (2) apparently doesn't carry a volume number, so the year should precede the page number;
– an "et al." is inappropriate after the name of a first author (7);
– both the year and place of publication are missing from the book citation in (3), and the same is true for (4)—assuming this *is* referring to a book [too little information is provided to tell, and if it is, title-word capitalization is inconsistent between (3, 4)];

– lecture notes (5) would not be classified as part of "the (published) literature", and would therefore warrant mention at most in a footnote;

– item numbers in a list should not be followed by *both* a period and a close parentheses sign; with items in a bibliographic list it is best to employ sets of parentheses (better: square brackets), or to present the numbers as superscripts.

Corrected version, with (4, 5) deleted:

[1] Meier P. Fresenius Z Anal Chem. 1987; 245: 211.
[2] Müller HP, Hausbold R. Proc Royal Chem Soc. London. 1980: 134.
[3] Meyer VR. Practical High-Performance Liquid Chromatography. 4th ed. New York: John Wiley; 2004.
[4] Princeling M., M.S. Thesis, Georgia Institute of Technology; 2001.
[5] Pákányi L, Bihácsi L, Hencsel P. Cryst Struct Commun. 1978; 7: 435–442.

S 15–2   Problematic features:

– Year identifications positioned inconsistently, sometimes with and sometimes without parentheses;

– authors' initials appear randomly before or after last names (and in one case not at all);

– inconsistent treatment of journal titles with respect to abbreviation;

– improper sequence for the references.

Corrected version:

Bard Y. 1974. Nonlinear Parameter Estimation. New York: Academic Press. p. 145.
Milow M. 1980. Talanta. 1037–1044.
Milow M. 1984a. Talanta. 1083–1087.
Milow M. 1984b. Inorg Chim Acta. 26: 947–951.
Nagano K, Metzler F. 1967. J Amer Chem Soc. 89: 2891.
Polster J. 1975. Z Phys Chem N F. 97: 113–118.
Ricci JE. 1952. Hydrogen Ion Concentrations. Princeton, NJ: Princeton University Press.

S 15–3   The citations should appear in the following order:

f, j, b, g, a, d, e, c, i, h

Also, the year of publication should be supplemented for **f** and **j** (1968a, 1968b). The order of these two might need to be reversed as well, depending upon where each is first mentioned in the text.

S 15–4   Among various additional possibilities are corporate publications, progress reports (and reports in general), grant applications, legal statutes, standards, and patents—and electronic resources of all kinds (e.g., CD-ROMs, Web sites, etc.).

15

S 15–5 Not really. Not mentioned, for example, are books with multiple editions, books with editors, multivolume works, and individual chapters within books.

S 15–6 No, it is much better insofar as possible to point the reader to the precise location of relevant information; for example:

... ; chap 5.
... ; 1992: 112–113.

S 15–7 Smith H, Johnson MF. In: Shulz B, ed. Insects of the Amazon Rain Forest; vol 20. Tallahassee/Florida: World Press; 1990: 107–117.

S 15–8a ... became accessible through a "half-sandwich" structure;[45] apart from HX, numerous other electrophiles containing sulfur,[46] selenium[47] and tellurium[48–50] as the key atom—the same applies as well to carbenes[51,52] and nitrenes[53] — are subject to addition, as are certain metal compounds with Lewis-acid character, including CuCl, among others.[54]

S 15–8b ... Müller [12], Kandroro [13–16], Finnigan [17] and also Mertz et al. [18] have established that the process in fact consists of an acid-catalyzed isomerization of **20** to **21**; complexes with linear [19] building blocks X—especially spectacular: X = H–C≡C–H as ligand [20]—have also been prepared quite recently [21–22]; experiments with Y–C≡C–H (Y = Me) [23] were unsuccessful [24].

Unit 16 Solutions

S 16–1 Appendix A. Microtome Thin Sections

(with A.1 through A.n for the individual photomicrographs)

Appendix B. Tables of Measured Coordinates

(with B.1 through B.n for the various tables)

Appendix C. Computer Programs

S 16–2 Parenthetical inserts in the text, footnotes, inclusion of additional information with the references, appendices.

S 16–3 (Individualized, specific response.)

Unit 17 Solutions

S 17–1 The information in question might be collected in an appendix entitled "Remarks", or combined with the bibliography to form a "Literature and Remarks" section.

S 17–2 In running text: superscripts [1], [2] etc., *), ‡), †) etc., as well as *), **); in tables, (see also Unit 20) preferably superscript letters [a], [b] etc.

S 17–3 The two types of superscript would obviously be subject to misinterpretation. In the absence of superscript citation numbers—i.e., if citation numbers appear on the text baseline, in either parentheses or square brackets, or when the name–date reference system has been employed—it is perfectly acceptable to indicate footnotes by superscripted numbers. Otherwise, special symbols like [*] or [+] must be utilized.

S 17–4 See Ex 17–4a and Ex 17–4b.

S 17–5 Text text text text.[1] Text text text text text text,[2] text text text text,[3,4] text text;[5] text text text text text text,[6–9] text text—text text text[10]—text text text text text text.[11]

---

[1] Footnote footnote footnote footnote footnote footnote footnote …

Unit 18 Solutions

S 18–1 Certainly not! In the first place, it is not "equations" one is supposed to italicize, but specific, individual symbols (those for quantities, for example). Moreover, in the absence of an italic font (e.g., for a typewritten document), it may suffice to underline those symbols that should be italicized.

S 18–2 Roman.

S 18–3 In text, write out the complete word (e.g., "several centimeters"), but in conjunction with numbers use the appropriate symbol (e.g., 2.5 cm).

S 18–4 Unlike "general functions", which are italicized, "special functions" (like "sin", "cos", "log", or "ln") are set in roman type.

S 18–5 The most important exception to the rule that numbers should always appear in roman (upright) type is a number used to designate some feature in a figure (which should instead be italicized).

S 18–6 Vectors and matrices are distinguished by being set in boldface italic type. With a typewritten manuscript, vector notation should adhere to the traditional pattern of a symbol surmounted by an arrow (introduced by hand).

S 18–7 Both are set in roman type. They differ, however, in the fact that "ppm" is preceded by a blank space, whereas the differential operator is set directly adjacent to an expression to which it applies.

S 18–8 Ideally, indices (subscripts, superscripts) should be made roughly 2/3 the size of the host symbol.

S 18–9 Yes, they should be italicized.

18

S 18–10   A single blank space is always to be inserted between a number and the corresponding unit. In the case of the percent sign, the absence of such a space (or insertion of at most a very thin space) is quite common.

S 18–11   12 mol : L is definitely unacceptable; 12 mol · L^{-1} would class as highly unusual (the dot should be eliminated), and 12 · mol/L is in violation of the rules.

S 18–12   The prefixes "d" and "h" are to be avoided, use of multiple prefixes (mμL) is forbidden, and it is recommended that prefixes be so adjusted that numerical values fall between 0.1 and 1000. Correct notation here would thus be:

0.2 m; 12 nL; 70 μmol; 2.450 m; 89.5 kPa; 12.80 μA

S 18–13   A useful rule of thumb is to supply a list whenever more than 10 seldom-encountered symbols are present, and also if there is any possibility of ambiguity. Often a combined "List of Symbols and Abbreviations" is most appropriate. The proper place for such a list is the page prior to the start of the actual text of a thesis.

S 18–14   There is inconsistency in the way units are expressed (e.g., both N/mm^2 and N mm^{-2}), all quantity symbols should be italicized, and two instances of wrong usage are present (units set in square brackets, absence of parentheses in the combination g/cm · h · mbar). Also note that the definitions of δ_0 and P do not in fact entail mathematical equations, so the corresponding equals signs should be eliminated. Finally, no unit is specified for the last entry. Better:

a_i impact strength, kJ/m^2
δ_0 bulk density, g/cm^3
H spherical indentation hardness, N/mm^2
P mean surface pressure, N/mm^2
P_{WD} permeation coefficient for steam, g/(cm · h · mbar)

Unit 19 Solutions

S 19–1   A standard keyboard provides access to only one character for representing three distinct symbols: the minus sign, the hyphen, and the "long dash" that signifies a pause. Nevertheless, a word processor makes it possible to represent each in a unique way (i.e., correctly!). In particular, you should strictly avoid approximating a minus sign by the much shorter hyphen.

S 19–2   Fractions can be indicated either with a slash (/) or using the traditional "horizontal line" notation. A colon should never be called upon for this purpose, however.

S 19–3a   The absence here of blank spaces before and after symbols like "+", "=", or ">" is especially disturbing. Moreover, the letters "k", "c", "i", and "n" should all be italicized, and a true minus sign should be used in the superscript. Better:

… produces consistent with the relationship

$$k_1 = k_n c_2 n^{-1} + k_0 \tag{22}$$

and in the special case of $n > 3$:

$$k_1 = k_2 c_i + k_0 \ (i = 1, \ldots, n) \tag{23}$$

S 19–3b   The order in which "fences" have been introduced is incorrect, and two fence styles will actually suffice. Italics are also missing for "x" and "y", and the minus signs have been inappropriately represented by hyphens. Better:

$$y = 3\,[x - 2\,(x + 3) \cdot (x^2 - 2x + 14)]$$

S 19–3c   The integral and summation signs are too small.

S 19–4a   The first line must not be allowed to end in the middle of an equation; the first term of the latter (x_{AWS}) must therefore be moved to the next line. Note also the inconsistency in symbolic representation of the "liter" ("l" vs. "L"). Increasingly, only the capitalized form is considered proper.

S 19–4b   Equations should be uniformly indented, not set flush with the left margin. On the other hand, equation numbers *do* belong next to the margin: the *right* margin.

Unit 20 Solutions

S 20–1a   **Table 11**. Text for the table caption.

x	y	x	y
1	123	25	84
5	118	40	315
10	22		

S 20–1b   **Table 4–2**. Time-dependence of the concentration c of the solution.

t	c	t	c
min	mol/L	min	mol/L
0.5	$1.526 \cdot 10^{-6}$	15	$4.728 \cdot 10^{-5}$
2	$7.015 \cdot 10^{-6}$	19.5	$1.033 \cdot 10^{-4}$
12.5	$0.261 \cdot 10^{-5}$	42	$5.052 \cdot 10^{-4}$

S 20–1c   **Table 4–4.** Yields in the reaction of CH_3SO_2Cl with H_2NOH.

Temperature, °C	50	60	70	90	100	110	120
Yield, %	65	68	75	92	95	95	59[a]

[a] partial decomposition

S 20–1d   **Table 12.** Properties of several solvents.

	Molar mass (in g/mol)	m.p. (in °C)	b.p.[a] (in °C)	Density (in g/cm³)
Benzene	78.12	5.5	80.1	0.87865
Phenol	94.11	43	181.75	1.0722
Toluol	92.15	–95	110.6	0.8669

[a] at 760 mbar

S 20–2   Here one could dispense with a table altogether. A possible substitute in the form of text might then read:

… The level of **13** was measured at two smokestack locations (measurement points *1* and *2*; flow rate 2.5 m³/min and 3.0 m³/min, respectively; exhaust gas temperature 604…606 °C) at various times ranging between 5 min and 1 h after feed. It varied only slightly around a mean value of 7 ppm, thus falling significantly below the 50-ppm limit prescribed by the federal government [12].

If a table were still considered necessary, the one suggested should certainly be modified; e.g.:

Table 2–1. Level of **13** in flue gas. – Measurement point 1, flow rate 2.5 m³/min.

Time, min	Temperature, °C	Content, ppm
5	606	8
10	604	6
15	605	7
20	606	6
30	604	7
60	608	7
30	606	7[a]
60	606	7[a]

[a] measurement point 2, flow rate 3.0 m³/min

S 20–3   Melting points and refractive indices for compounds of the type $CH_3-SO_2-N(R^1)OR^2$ have been collected in Table 3.

Table 3. Melting points and refractive indices for compounds of the type $CH_3-SO_2-N(R^1)OR^2$ [3].

R^1	R^2	Melting point °C	n_D^{20}
H	H	172	
H	CH_3	132	1.3528
CH_3	H	145	1.4255
CH_3	CH_3	152[a)]	1.4352

[a] with decomposition

S 20–4   After hydrolysis of **3** using water that had been enriched in ^{18}O, the observed intensity ratio for peaks at $m/e = 46$ and $m/e = 44$ in the mass spectrum of released dinitrogen monoxide was $(2.0 \pm 0.5) \cdot 10^{-3}$, whereas with non-enriched water the corresponding ratio was $(2.5 \pm 1.2) \cdot 10^{-3}$.

S 20–5   Headings **a** and **c** are both acceptable. In the case of **b**, however, units should not be set in square brackets, and with **d** there should be consistent use of either quantity names or quantity symbols.

S 20–6   All vertical lines should be eliminated, as should the three horizontal lines within the body of the table.

Unit 21 Solutions

S 21–1   "Anchoring a figure" means making explicit reference to that figure, by number, somewhere in a document's running text. All figures are expected to be anchored in this way.

S 21–2   Every figure is provided with a caption that describes its content. Often such a caption is supplemented—at the end, in a clearly delineated way—by additional explanatory information, referred to as a "legend" (cf. Ex 21–10).

S 21–3   Through limiting any text within the figure to symbols and abbreviations, which are then explained in a figure legend added at the end of the caption.

S 21–4   Halftone copy (unlike line drawings) is characterized by the presence not only of black and white areas, but also gray shades—potentially representing a true continuum. This feature makes it much more difficult to prepare duplicates of halftones without serious loss in quality.

S 21–5   Lines present in a graph should be so constructed that their intensities (in effect, their widths) correspond approximately to their relative importance.

21

A common rule suggests that lines for curves, axes, and subsidiary elements like grid markings should have widths in the ratio 2 : 1.4 : 1.

S 21–6   Numbers: along axes, upright (roman), but for designating elements within a figure, italic; tick marks: all equally long and intense, and generally directed toward the interior of the graph; quantity symbols: italicized, preferably set below the abscissa and to the left of the ordinate, but located with qualitative graphs at the base of the directional arrows; arrows: sometimes incorporated into the axes themselves. Labels within a figure should be sized so that they are roughly consistent with type in the surrounding text.

S 21–7   Unit symbols should either be incorporated into the scale markings, or set after the appropriate quantity symbols (cf. Figures 21–8 and 21–5).

S 21–8a   This is a qualitative representation. Missing, however, are arrows indicating directions of increase for the two quantities. The addition of selective crosshatching or dignified background patterns would make it easier to distinguish the various regions of interest. The intensity of the lines for all curves should be roughly doubled, since these are of the utmost importance.

S 21–8b   Another qualitative representation. The symbols "A" and "B" should be set not as here at the ends of arrows, but to the left of and beneath the arrows, respectively. Moreover, the arrows themselves belong *outside* the graph.

S 21–8c   Tick marks should be directed inward, and on both axes too many (unlabeled) tick marks have been provided. Given the absence of units, "Mixing ratio" is too vague a term. The label "Height in km" should be rotated 90° counterclockwise and set as a single text line. Separate the scales more clearly, and show zero points for each.

S 21–8d   As with **8c**, separate the scales more clearly. Curve numbers should be italicized. The "h" and "L" would presumably be set in italics in the associated text, and if so, the same should be true in the figure.

S 21–8e   Explicit zeroes should be shown for both axes. Units should not be enclosed in square brackets. On the ordinate, it is not good form to display the factor "10^4" in conjunction with only one numerical value. Better: Use as a label "Light emission in 10^4 impulses", where the unit can either be presented in this fashion or in parentheses. On the abscissa, no quantity has been indicated. The applicable unit is simply "mg/mL" (without square brackets!), and the "SO_2" should in fact become a part of the quantity defi-

21

nition; e.g., "$c(SO_2)$" or "SO_2 concentration". Framing at the top and right-hand side of the graph should be eliminated.

S 21–8f   The data points are too small, and lines associated with them should be made more intense (thicker). On the abscissa, "Minutes" should be replaced by the quantity actually measured, with the corresponding unit appended to it (e.g., "Decomposition time, min"). On the ordinate, "A^{2+}" is *not* a quantity. What is presumably meant is the *concentration* of that species: "$c(A^{2+})$".

S 21–8g   Scales have again not been clearly separated. It would be preferable to move the abscissa scale downward a bit so that the two axes, which do not have a common "origin", do not cross at all. Arrows along calibrated axes are superfluous, and the axis labels should be centered. For better visibility, the percent sign should be rotated 90° clockwise and set between the scale markings "1" and "2".

S 21–8h   All tick marks should be the same length, and directed inward. Numerical values along the ordinate should be rotated 90° clockwise. The ordinate has been subdivided too finely; steps of 0.1 or even 0.2 units would suffice. Within the diagram, it would look better if the numerical values were set *inside* the bars, near the top, and in smaller print. Year information should be provided *below* the abscissa, not within the bars.

21

Index